Francis Orpen Morris

A Natural History of British Moths

Volume II

Francis Orpen Morris

A Natural History of British Moths
Volume II

ISBN/EAN: 9783744790994

Printed in Europe, USA, Canada, Australia, Japan

Cover: Foto ©berggeist007 / pixelio.de

More available books at **www.hansebooks.com**

BRITISH MOTHS,

ACCURATELY DELINEATING EVERY KNOWN SPECIES,
WITH THE ENGLISH AS WELL AS SCIENTIFIC NAMES, ACCOMPANIED BY FULL DESCRIPTIONS,
DATE OF APPEARANCE, LISTS OF THE LOCALITIES THEY HAUNT,
THEIR FOOD IN THE CATERPILLAR STATE, AND OTHER FEATURES OF THEIR HABITS
AND MODES OF EXISTENCE, ETC., ETC.

BY

THE REV. F. O. MORRIS, B.A.,

Author of A History of British Birds, A History of the Nests and Eggs of British Birds,
A Natural History of British Butterflies,
Etc., Etc., Etc.

THE PLATES CONTAIN NEARLY TWO THOUSAND EXQUISITELY COLOURED SPECIMENS.

COMPLETE IN FOUR VOLUMES.

VOLUME II.

LONDON:
HENRY EDWARD KNOX.
1871.

CONTENTS OF VOL. II.

	Page
LARENTIDÆ (continued).	
Collix	1
Lobophora	1
Thera	4
Ysipetes	7
Melanthia	8
Melanippe	10
Anticlea	15
Coremia	18
Camptogramma	20
Phibalapteryx	21
Scotosia	24
Cidaria	27
Coremia	31
Pelurga	35
Eubolia	36
Carsia	39
Anaitis	40
Lithostege	40
Chesias	41
Tanagra	42
DREPANULÆ.	
Platypteryx	42
Drepanula	43
Cilix	46
DICRANURIDÆ.	
Cerura	47
Stauropus	50
Petasia	50
PSEUDO-BOMBYCES.—PYGÆRIDÆ.	
Pygæra	51
Clostera	52
NOTODONTIDÆ.	
Gluphisia	54

	Page
Ptilophora	54
Ptilodontis	55
Notodonta	56
Diloba	62
TRIFIDÆ.—BOMBYCIFORMES.	
Thyatira	63
Noctua	64
Cymatophora	64
BRYOPHILLIDÆ.	
Bryophila	68
BOMBYCOIDÆ.	
Dipthera	70
Acronycta	71
NOCTUÆ.	
Simyra	79
GENUINÆ.—LEUCANIDÆ.	
Synia	79
Leucania	80
Meliana	87
Senta	87
Nonagria	88
APAMIDÆ.	
Gortyna	92
Hydrœcia	93
Axylia	95
Xylophasia	96
Dipterygia	99
Xylomiges	100
Aporophila	100
Laphygma	101
Neuria	101
Heliophobus	102
Charæas	103
Pachetra	104

CONTENTS.

	Page.		Page.
Cerigo	104	Pachnobia	151
Luperina	105	Tæniocampa	152
Crymodes	107	Orthosia	159
Mamestra	107	Anchocelis	162
Apamea	111	Glæa	165
Miana	115	Scopelosema	167
Celæna	117	Dasycampa	167
CARADRINIDÆ.		Hoporina	168
Grammesia	118	Xanthia	169
Hydrilla	119	Cirrædia	172
Acostemetia	119	COSMIDÆ.	
Caradrina	120	Tethea	173
NOCTUIDÆ.		Euperia	174
Rusina	122	Dicycla	175
Agrotis	123	Cosmia	175
Triphæna	135	HADENIDÆ.	
Noctua	139	Eremobia	178
Trachea	151	Dianthæcia	179

NATURAL HISTORY OF BRITISH MOTHS.

COLLIX SPARSARIA.

Plate XXXI. *Figure* 1.

Localities for this species are York, the New Forest, Bromborough, Birkenhead, Cambridge.

The perfect insect appears in June.

The caterpillar is pale green, with five white lines along the back, and a broad yellow side line.

It feeds on the yellow loosestrife (*Lysimachia vulgaris*).

I have to return my best thanks to the Rev. R. P. Alington, M.A., Rector of Swinhope, Lincolnshire, for valuable assistance in this volume.

LOBOPHORA SEXALARIA.

SMALL SERAPHIM.

Plate XXXI. *Figure* 2.

Localities for this species are York, Brighton, Croxteth, Cambridge, Halton, Ambleside, Keswick, Lewes, Manchester, St. Osyths, Lewisham, Kingsbury, Worthing, Lyndhurst, Stowmarket.

The situations where it is found are woods.

The perfect insect appears in June, July, and August

The caterpillar is pale whitish-green, with a white line along the back, another on each side below it, and the side line the same colour; the head dark green; the last segment has two small projections.

The date of the appearance of the caterpillar is in August and September.

It feeds on the sallow, the willow, &c.

The chrysalis is subterranean.

My best thanks to the Rev. H. A. Stowell, of Christ Church Parsonage, Maughold, Isle of Man, for great assistance in this work generally.

LOBOPHORA HEXAPTERARIA.

Plate XXXI. *Figure* 3.

Localities for this species are Scarborough, Rotherham, Brighton, Sheffield, Swanscombe, Worcester, Cambridge, Bristol, Kingsbury, Lewes, Stowmarket, Newcastle-on-Tyne.

The perfect insect appears in June.

The caterpillar is bright green, with a yellow line along both sides, and two yellow points on the hind segment; two also on the head.

The date of the appearance of the caterpillar is in June.

It feeds on the sallow and the aspen.

The chrysalis is found under the earth.

My thanks are due to Mr. William Prest, of York, for many valuable aids and assistances.

LOBOPHORA VIRETARIA.

Plate XXXI. *Figure* 4.

Localities for this species are Brighton, Black Park, Worcester, Newcastle-on-Tyne, Lyndhurst, Lewes, Cambridge.

The perfect insect appears in May and June.

The caterpillar is dull yellow, spotted on the back with dull orange.

The date of the appearance of the caterpillar is in August and September.

It feeds on the privet?

The chrysalis is found beneath the earth.

To Mr. S. P. Saville, of Dover House, Cambridge, my best thanks are due.

LOBOPHORA LOBULARIA.

THE EARLY TOOTH-STRIPED.

Plate XXXI. *Figure* 5.

Localities for this species are York, Brighton, Kirkby near Liverpool, West Wickham, Dulwich, Durham, Manchester, Newcastle-on-Tyne, Cambridge, Lewes, Ipswich, Newnham, Exeter, Edinburgh, Gourock, Dunoon, Bristol.

The perfect insect appears in April.

The caterpillar is dark green, with a broad yellowish side line.

The date of the appearance of the caterpillar is in August.

It feeds on the sallow.

The chrysalis is found beneath the surface of the earth.

I have to record my thanks to Mr. Robert Anderson, of York, for several useful communications.

LOBOPHORA POLYCOMMARIA.

SPRING CARPET.

Plate XXXI. *Figure* 6.

Localities for this species are Brighton, Dartford, Lewes.

The perfect insect appears in April.

The caterpillar is dark green, the sides paler; the side line pale yellow.

The date of the appearance of the caterpillar is in August.

It feeds on the honeysuckle.

The chrysalis is subterranean.

I am indebted also for communications to Mr. Edward Gleave, of Birkenhead.

THERA JUNIPERARIA.

JUNIPER CARPET.

Plate XXXI. *Figure* 7.

Localities for this species are Barnstaple, Worcester, Arran, Sanderstead, Mickleham, Glasgow, Box Hill.

The perfect insect appears in October.

The caterpillar is pale green, whitish on the upper part, the sides darker, with a broad pale yellow line on either side below the back; side line dark purple red, edged on the lower side with white.

The date of the appearance of the caterpillar is in July and August.

It feeds on the juniper.

The chrysalis is placed among leaves in a silken cocoon.

I have likewise here to thank Mr. W. F. Kirby, of St. Peter's House, Brighton.

THERA SIMULARIA.

BRICK CARPET.

Plate XXXI. *Figure* 8.

Localities for this species are Perry Wood near Faversham, Isle of Man, Darlington, Flesk, West Wickham, Goldspie, Sutherland, Edinburgh, Glasgow, and Newcastle-upon-Tyne.

The situations where it is found are fir plantations.

The perfect insect appears in July and October?

The caterpillar is bright green, with a bluish white line along the back, a white line on each side of it; the side line white edged with red on its upper side.

The date of the appearance of the caterpillar is in June.

It feeds on the juniper.

The chrysalis is found among leaves in a cocoon of silk.

I desire here to thank Mr. John Birks, of York, for his assistance in this volume.

THERA VARIARIA.
GREY CARPET.
Plate XXXI. *Figure* 9.

Localities for this abundant species are York, Brighton, West Wickham, Barnstaple, Arran, Black Park, Faversham, Isle of Man.

The perfect insect appears in June and September.

The caterpillar is green, with a white line along the back; the side line white.

The date of the appearance of the caterpillar is in March, May, and July.

It feeds on the fir.

The chrysalis is found among leaves in a cocoon of silk.

Here I must also thank Mr. J. H. Dossor, of York.

THERA FIRMARIA.
Plate XXXI. *Figure* 10.

Localities for this species are York, Scarborough, Brighton, Wavendon, Paisley, Lyndhurst, Fort Augustus, Edinburgh, Sutherland, Manchester, Inverness, Golspie, Pembury, Exeter, Isle of Man, Ipswich, Birkenhead.

The perfect insect appears in July.

The caterpillar is dull green; the side line yellow, head brown.

The date of the appearance of the caterpillar is in April.

It feeds on the fir.

The chrysalis is found in a cocoon of silk among leaves.

I take this opportunity of thanking the Rev. G. Rudston Read, Rector of Sutton-on-Derwent, for much useful information.

YPSIPETES RUBERARIA.

Plate XXXI. *Figure* 11.

Localities for this species are York, Wimbledon, Birkenhead, Barnstaple, and in the Fens of Cambridgeshire and Huntingdonshire.

The situations where it is found are the fens.
The perfect insect appears in May and June.
The caterpillar is dull white or greyish.
It feeds on the sallow and the willow.
The chrysalis is found among leaves.

YPSIPETES IMPLUVIARIA.

MAY HIGHFLIER.

Plate XXXI. *Figure* 12.

Localities for this very plentiful species are York, Fairbrook, Faversham, Brighton, Kilmun, Scarborough, Carron, Stirling, Barnstaple, Darlington, Huddersfield, Birkenhead, Edinburgh, Lyndhurst, Brighton, Exeter, Halton, Manchester, Cambridge, Newcastle-upon-Tyne, Pembury, Lower Guiting, Stowmarket.

The situations where it is found are woods.

The perfect insect appears in May.

The caterpillar is yellowish, with a bluish-green line along the back, and another of the like colour on each side of it; the side line the same.

The date of the appearance of the caterpillar is in September.

It feeds on the alder and the geranium.

The chrysalis is found among leaves.

YPSIPETES ELUTARIA.

JULY HIGHFLIER.

Plate XXXI. *Figure* 13.

Localities for this very common species are York, Nunburnholme, Brighton, Humberstone, Faversham, Isle of Wight, Barnstaple, Arran, Falmouth, Bromsgrove, Charmouth.

The situations where it is found are woods.

The perfect insect appears in July and August.

The caterpillar is dull yellow, with a pale yellow line on each side below the back, and under it two smaller ones: the side line yellow, with a reddish-brown spot in the middle of each segment.

The date of the appearance of the caterpillar is in May and June.

It feeds on the sallow, the alder, the bilberry, &c.

The chrysalis is found among leaves.

MELANTHIA RUBIGINARIA.

BLUE BORDERED CARPET.

Plate XXXI. *Figure* 14.

Localities for this species are York, Sheffield, Faversham, Scarborough, Brighton, Simonswood near Liverpool, Halton, Preston, Carlisle, Cambridge, Stowmarket, Lyndhurst, Kingsbury, Tenterden, Glasgow, Barnstaple, Newnham, Lower Guiting, Darlington, Edinburgh, Exeter, Manchester, Newcastle-on-Tyne.

The situations where it is found are woods.

The perfect insect appears in June, July, and August.

The caterpillar is pale green, with a dark green line along the back, and a line on each side beneath it of yellowish-green.

The date of the appearance of the caterpillar is in June. It feeds on the alder.

The chrysalis is found among leaves or in a cocoon of earth.

MELANTHIA OCELLARIA.

PURPLE BAR.

Plate XXXI. *Figure 15.*

Localities for this abundant species are York, Brighton, Humberstone, Faversham, Exeter, Bristol, Isle of Man, Glasgow.

The situations where it is found are woods.

The perfect insect appears in June and July.

The caterpillar is brown, with white lines on the sides.

The date of the appearance of the caterpillar is in June and September.

It feeds on the bedstraw *(Galium verum)*.

The chrysalis is found between leaves, or in a cocoon of earth.

MELANTHIA ALBICILLARIA.

Plate XXXII. *Figure 1.*

Localities for this species are York, Rotherham, Scarborough, Blean Woods near Canterbury, Fairbrook, Faversham, Brighton, Hale near Liverpool, Manchester, Black

Park, Ambleside, Cambridge, Halton, Dorking, Killarney, Pembury, Plymouth, Chatmoss, Lewes, Lyndhurst, Preston, Newnham, Newcastle-on-Tyne, Darlington, Bristol, Whalley, Tenterden, Barnstaple.

The situations where it is found are woods.

The perfect insect appears in June and July.

The caterpillar is green, with triangular-shaped spots along the back on the fourth, fifth, sixth, seventh, eighth, ninth, and tenth segments; the side line white.

The date of the appearance of the caterpillar is in September and October.

It feeds on the bramble and the raspberry.

The chrysalis is found among leaves, or in a cocoon of earth.

MELANIPPE HASTARIA.

ARGENT AND SABLE.

Plate XXXII. *Figure* 2.

Localities for this species are York, Huddersfield, Scarborough, Buttercrambe Moor near Stamford Bridge, Yorkshire; the Ran-Dan woods and Cobbler's coppice near Bromsgrove, Lewes, Linwood near Market Rasen, Ben Lomond, Dunoon, Arran, Tenterden, Stowmarket, Manchester, Lyndhurst, Newcastle-upon-Tyne, Newnham, Witney, Carlisle, Worcester, Sudbury, Lower Guiting, and Killarney.

The situations where it is found are woods.

The perfect insect appears in May.

The caterpillar is brown or blackish-brown, with a dark brown line along the back.

The date of the appearance of the caterpillar is in August.

It feeds on the birch, and is gregarious.

The chrysalis is enclosed in a cocoon of earth.

MELANIPPE TRISTARIA.

SMALL ARGENT AND SABLE.

Plate XXXII. *Figure* 3.

Localities for this species are Huddersfield, Edinburgh, Stowmarket, Newnham, Newcastle-on-Tyne, Sheffield, Galway, Kilmun, Torwood, Ben Nevis.

The perfect insect appears in June and July.

The caterpillar is dull yellowish-brown, with a brown line along the back, and the side line brown also.

The date of the appearance of the caterpillar is in August and September.

It feeds on the bedstraw (*Galium verum*).

The chrysalis is enclosed in an earthen cocoon.

MELANIPPE PROCELLARIA.

CHALK CARPET.

Plate XXXII. *Figure* 4.

Localities for this species are Brighton, Faversham, Canterbury, Sanderstead, Plumstead, Bristol, Dorking, Cambridge, Black Park, Ipswich, Lewes, Poynings, Ventnor, Malvern, Newcastle-on-Tyne.

The situations where it is found are woods.

The perfect insect appears in June and July.

MELANIPPE UNANGULARIA.

SHARP-ANGLED CARPET.

Plate XXXII. Figure 5.

Localities for this species are Brighton, Birkenhead, Barnstaple, West Looe, Preston, Exeter, Ipswich, Lewes, Pembury, Stowmarket, Manchester, Newcastle-on-Tyne.
The situations where it is found are woods.
The perfect insect appears in June and July.

MELANIPPE RIVARIA.

WOOD CARPET.

Plate XXXII. Figure 6.

Localities for this species are Brighton, Faversham, West Looe, Lewes, Worcester, Bristol, Preston, Ipswich, Barnstaple, Pembury, Manchester, Tenterden.
The situations where it is found are chalky places.
The perfect insect appears in July.
The caterpillar is greenish-grey, the back greener, with a row of dark green lozenge-shaped spots, edged with white on the fifth, sixth, seventh, eighth, and ninth segments; the side line yellowish.
The date of the appearance of the caterpillar is in September.
It feeds on the yellow bedstraw (*Galium verum*).
The chrysalis is enclosed in a cocoon of earth.

MELANIPPE SUBTRISTARIA.

COMMON CARPET.

Plate XXXII. Figure 7.

Localities for this species are York, Brighton, Falmouth.
The situations where it is found are woods and hedges.
The perfect insect appears in May and August.
The chrysalis is enclosed in a cocoon of earth.

MELANIPPE MONTANARIA.

SILVER-GROUND CARPET.

Plate XXXII. Figure 8.

Localities for this very common species are York, Brighton, Humberstone.

The situations where it is found are hedge-sides, woods, etc.

The perfect insect appears in May and August.

The caterpillar is dull whitish, with several greyish-brown lines along it.

The date of the appearance of the caterpillar is April and May.

It feeds on the primrose *(Primula vulgaris.)*

The chrysalis is enclosed in an earthen cocoon.

MELANIPPE GALIARIA.

GALIUM CARPET.

Plate XXXII. *Figure* 9.

Localities for this species are York, Huddersfield, Scarborough, Brighton, Humberstone, Dublin, Lewes, Weston-on-the-Green, Birkenhead, Lyndhurst, Arundel, Bristol, Darlington, Ventnor, Edinburgh, Manchester, Plymouth, Portland, Exeter, Newcastle-on-Tyne, Lynton, West Looe, and the Cumbraes in Scotland.

The perfect insect appears in May, June, July, and August.

The caterpillar is greyish-brown, with a black line along the back, and another white one tinged with rose colour below it on each side, the side line reddish-grey.

The date of the appearance of the caterpillar is in July.

It feeds on the bedstraw (*Galium verum.*)

The chrysalis is enclosed in a cocoon of earth.

This species has been known with one hind wing, and even both wanting.

MELANIPPE FLUCTUARIA.

GARDEN CARPET.

Plate XXXII. *Figure* 10.

Localities for this species are York, Brighton, Humberstone, Falmouth, and near London.

The situations where it is found are gardens, etc.

The perfect insect appears in April and August.

The caterpillar is yellowish-grey, mottled on the back with blackish-brown, the side line dull yellowish-grey.

The date of the appearance of the caterpillar is in June and September.

It feeds on the horse-radish, cabbage, &c.

The chrysalis is enclosed in a cocoon of earth.

ANTICLEA SINUARIA.

ROYAL MANTLE.

Plate XXXII. Figure 11.

Localities for this species are Exeter, West Looe, Cambridge.

The perfect insect appears in June.

The caterpillar is yellow on the back, with a broad line on each side below it of a purple colour, edged on the lower side with greenish-yellow.

The date of the appearance of the caterpillar is in July and August.

It feeds on the white bedstraw *(Galium verum)*.

The chrysalis is enclosed in a cocoon of earth.

ANTICLEA RUBIDARIA.

THE FLAME.

Plate XXXII. *Figure* 12.

Localities for this species are Brighton, Arundel, West Looe, Worcester, Bristol, Cambridge, Exeter, Manchester, Lewes, Newnham, Newcastle-on-Tyne, Pembury, Stowmarket, Tenterden.

The situations where it is found are woods.

The perfect insect appears in June and July.

The caterpillar is pale green or greyish, with a line along the back on the first, second, and three hind segments; the intermediate segments with a blackish network in which is a black triangular mark.

The date of the appearance of the caterpillar is August and September.

It feeds on the bedstraw *(Galium verum)*.

The chrysalis is found in a cocoon of earth.

ANTICLEA BADIARIA.

SHOULDER STRIPE.

Plate XXXII. *Figure* 13.

Localities for this very common species are York, Nunburnholme, Brighton, Bromsgrove, Falmouth, Wavendon, Epping, Manchester.

The perfect insect appears in March and April.

The caterpillar is green, mottled with blackish on the back; the side line pale green, with white spots; the head brownish-yellow.

The date of the appearance of the caterpillar is June.

It feeds on the rose.

The chrysalis is found in an earthen cocoon.

ANTICLEA DERIVARIA.

STREAMER.

Plate XXXII. *Figure* 14.

Localities for this species are York, Nunburnholme, Huddersfield, Scarborough, Brighton, Bromsgrove, Barnstaple, Darlington, Edinburgh, Kingsbury, Birkenhead, Exeter, Lewes, Lyndhurst, Worthing, Bristol, Newnham, Stowmarket, Cambridge, Newcastle-on-Tyne, Lewisham, Lower Guiting.

The situations where it is found are gardens.

The perfect insect appears in April, May, and June.

The caterpillar is pale green, with a red line on the first segments of the back, the segments divided with clear light yellow.

The date of the appearance of the caterpillar is in July.

It feeds on the rose and the honeysuckle.

The chrysalis is found in a cocoon of earth.

ANTICLEA BERBERARIA.

RASPBERRY CARPET.

Plate XXXII. *Figure* 15.

Localities for this species are Cambridge, Epping, Chelmsford.

The perfect insect appears in May and August.

The caterpillar is pale yellowish-brown, spotted on the back with black; the side line a pale dull yellowish-grey, with sometimes a tinge of rose colour.

The date of the appearance of the caterpillar is in June.

It feeds on the barberry.

The chrysalis is found in a cocoon of earth.

COREMIA MUNITARIA.

RUFOUS CARPET.

Plate XXXIII. *Figure* 1.

Localities for this species are Scarborough, Keswick, Darlington, Ben Nevis, Edinburgh, Arran, Luss, Inver, Manchester.

The perfect insect appears in June and July.

The chrysalis is found below the earth.

COREMIA PROPUGNARIA.

FLAME CARPET.

Plate XXXIII. *Figure* 2.

Localities for this species are York, Scarborough, Brighton, Barnstaple, Birkenhead, West Looe, Bristol, Cambridge, Kilmun, Darlington, Exeter, Edinburgh, Renfrew, Kingsbury, Lewes, Lyndhurst, Manchester, Newnham, Newcastle-on-Tyne, Pembury, Stowmarket, Tenterden.

The situations where it is found are woods.

The perfect insect appears in June and July.

The caterpillar is reddish-grey, with a row of pink triangular-shaped marks on the back; the side line greyish-yellow.

The date of the appearance of the caterpillar is September.

It feeds on the cabbage.

The chrysalis is found below the earth.

COREMIA FERRUGARIA.

RED TWIN-SPOT.

Plate XXXIII. *Figure* 3.

Localities for this species are York, Isle of Man, Humberstone, Falmouth, Stowmarket, Faversham, Birkenhead, Brighton, Halton, Tenterden, Barnstaple, Bristol, Lewes, Lyndhurst, Edinburgh, Cambridge, Manchester, Kingsbury, Glasgow, Exeter, Newnham, Worcester, Pembury.

The situations where it is found are hedge-sides.

The perfect insect appears in May, June, and August.

The caterpillar is greyish-brown, with an interrupted brownish line along the back, and a brown line on the sides.

The date of the appearance of the caterpillar is in June and September.

It feeds on the chickweed (*Alsine media*.)

The chrysalis is subterranean.

COREMIA UNIDENTARIA.

Plate XXXIII. *Figure* 4.

Localities for this species, which is closely allied to the preceding one, if indeed it be distinct from it, are York, Tenterden, Glasgow, Worthing, Barnstaple, Bristol, Cambridge, Birkenhead, Halton, Kingsbury, Brighton, Exeter, Lewes, Pembury, Manchester, Lyndhurst, Newnham.

The situations where it is found are hedge-sides.

The perfect insect appears in May, June, and August.

The caterpillar is greyish-brown, with an interrupted brownish line along the back, and a brown line on the sides.

The date of the appearance of the caterpillar is in June and September.

It feeds on the chickweed.

The chrysalis is subterranean.

COREMIA QUADRIFASCIARIA.

LARGE TWIN-SPOT.

Plate XXXIII. Figure 5.

Localities for this species are York, Cambridge, Stowmarket, Guildford, Perry Wood near Faversham, Worcester, Northleach, Black Park, Deal, Playford near Ipswich.

The situations where it is found are hawthorn hedges.

The perfect insect appears in June, at the end of the month, and in July.

The caterpillar is yellowish-grey mottled with brown, the side line blackish, sometimes interrupted.

The date of the appearance of the caterpillar is April, May, and August.

It feeds on the hawthorn and various low plants.

The chrysalis is found beneath the earth.

CAMPTOGRAMMA BILINEARIA.

YELLOW SHELL. ELM MOTH. ELM.

Plate XXXIII. Figure 6.

Localities for this extremely common species are York, Anstey, Brighton, Falmouth, Woodland near Broughton in Furness, Bromsgrove, Charmouth, Nunburnholme.

The situations where it is found are woods, hedges, gardens, etc.

The perfect insect appears in June and August.

The caterpillar is greenish-white, with a dark green line along the back, and a white line below it on each side; the side line whitish.

The date of the appearance of the caterpillar is in April, It feeds on grass, etc.
The chrysalis is found under the ground.

CAMPTOGRAMMA FLUVIARIA.

NARROW-BORDERED CARPET.

Plate XXXIII. Figure 7.

Localities for this species are Brighton, Birkenhead, Sidmouth, Hainault Forest, the Isle of Wight, Topsham, West Wickham, Dulwich, Exeter, Worthing, Barnstaple, Crosby in Lacashire, Regent's Park London.

The perfect insect appears in August and October, and on to January. August 5th, October 8th, January 1st.

The date of the appearance of the caterpillar is in August and September. September 7th.

It feeds on the spotted persicaria? (*Persicaria polygonum*).

The chrysalis is found under the earth.

PHIBALAPTERYX TERSARIA.

THE FERN MOTH.

Plate XXXIII. Figure 8.

Localities for this species are Brighton, Perry Wood near Faversham, Charlton, Darenth Wood, Arundel, Sedbergh, Barnstaple, Bristol, Cambridge, Ipswich, Lewes, Lyndhurst, Newnham, Stowmarket.

The situations where it is found are woods and hedges.
The perfect insect appears in June.

The caterpillar is pale brown, with several paler and darker lines along it, and mottled with brownish-black, with a brown line on the back edged with white; side line grey.

The date of the appearance of the caterpillar is in September and October.

It feeds on the clematis *(Clematis vitalba)*.

The chrysalis is found beneath the surface of the ground

PHIBALAPTERYX LAPIDARIA.

Plate XXXIII. *Figure 9.*

Localities for this species are Rannock, Perthshire.
The perfect insect appears in September.

PHIBALAPTERYX LIGNARIA.

OBLIQUE CARPET.

Plate XXXIII. *Figure 10.*

Localities for this species are York, Chilham, Battersea Fields, Hammersmith, Cambridge, Barnstaple, Birkenhead, Bristol, Edinburgh, Scarborough, Stowmarket.

The perfect insect appears in June and August.

PHIBALAPTERYX POLYGRAMMARIA.

Plate XXXIII. *Figure* 11.

Localities for this species are Bristol, Cambridge. The perfect insect appears in April and August.

PHIBALAPTERYX VITALBARIA.

SMALL WAVED-UMBER.

Plate XXXIII. *Figure* 12.

Localities for this species are Brighton, Sanderstead, Bristol, Cambridge, Dorking, Ipswich, Lewes, Arundel, Worcester, Worthing, Stowe Wood, Sudbury, Milstead.

The situations where it is found are woods and hedge sides.

The perfect insect appears in June.

The caterpillar is reddish grey, mottled with black, a dull yellowish red line on the side carried on to the hind segments; a black line on the back, and another on the sides, also black.

The date of the appearance of the caterpillar is in June and October.

It feeds on the clematis *(Clematis vitalba.)*

The chrysalis is subterranean.

SCOTOSIA DUBITARIA.

COMMON TISSUE.

Plate XXXIII. *Figure* 13.

Localities for this species are York, Brighton, Ilfracombe, Barnstaple, Birkenhead, Bristol, Cambridge, Lower Guiting, Darlington, Edinburgh, Exeter, Halton, Lewes, Huddersfield, Kingsbury, Lyndhurst, Manchester, Newnham, Newcastle-on-Tyne, Tenterden, Worthing.

The situations where it is found are gardens and woods.

The perfect insect appears in April and May, August and September.

The caterpillar is pale green, with four thin lines of white on the back; side line bright yellow.

The date of the appearance of the caterpillar is in July? It feeds on the buckthorn *(Rhamnus catharticus.)* The chrysalis is found under the ground.

SCOTOSIA VETULARIA.

BROWN SCALLOP.

Plate XXXIII. *Figure* 14.

Localities for this species are Brighton, Bristol, Cambridge, Sanderstead, Halton, Kingsbury, Worcester, Lewes, Newnham.

The situations where it is found are chalky places.

The perfect insect appears in June.

The caterpillar is bluish-grey, with two white lines on the back; the side line yellow and broad.

The date of the appearance of the caterpillar is in May.

It feeds on the buckthorn (*Rhamnus catharticus*).

The chrysalis is found below the ground.

SCOTOSIA RHAMNARIA.

DARK UMBER.

Plate XXXIII. *Figure* 15.

Localities for this species are York, Brighton, Bristol, Cambridge, Sanderstead, Halton, Lewes, Newnham, Wavendon, Arundel, Stowmarket, Worcester.

The situations where it is found are woods.

The perfect insect appears in May and July.

The date of the appearance of the caterpillar is in May.

The caterpillar is dark brown on the back, white mottled with brown on the sides; a variety is green, with a white line, edged below with dark reddish-brown, on the side.

It feeds on the buckthorn (*Rhamnus catharticus*).

The chrysalis is subterranean.

SCOTOSIA CERTARIA.

SCARCE TISSUE.

Plate XXXIV. *Figure* 1.

Localities for this species are Worcester, Newnham, Bristol, Cambridge.

The situations where it is found are gardens.

The perfect insect appears in April and May.

The caterpillar is purple-grey, with a darker line along the back, and another also darker on each side below it; the side line pale grey, with an orange spot on each segment.

The date of the appearance of the caterpillar is in June.

It feeds on the barberry.

The chrysalis is found below the earth.

SCOTOSIA UNDULARIA.

SCOLLOP-SHELL.

Plate XXXIV. *Figure 2.*

Localities for this species are Langwith near York, Brighton, Eastham near Birkenhead, Linwood near Market Rasen, Bristol, Arundel, Exeter, Halton, Lyndhurst, West Looe, Manchester, Newnham, Worcester, Pembury, Plymouth, Northleach, Stowmarket, Tenterden, Preston, Worthing, and Oxford.

The situations where it is found are woods.

The perfect insect appears in June.

The caterpillar is blackish-grey, with two narrow dusky white lines on the back, and another similar one below it, only less distinct; the side line dusky white and broad.

The date of the appearance of the caterpillar is in September and October.

It feeds on the sallow (*Salix capræa*).

The chrysalis is found under the ground.

CIDARIA PSITTACARIA.

RED-AND-GREEN CARPET.

Plate XXXIV. Figure 3.

Localities for this species are York, Brighton, Eastham near Birkenhead, Plymouth, Norbury Park, Bristol, Halton, Darlington, Scarborough, Barnstaple, Edinburgh, Exeter, Stowmarket, Worcester, Lyndhurst, Manchester, Charmouth, Newcastle-on-Tyne, Dunoon, and Innellan.

The situations where it is found are woods.

The perfect insect appears in September and October.

The caterpillar is greenish-yellow, beneath darker, with two red points on the hind segment, and sometimes a row of red spots on the back.

The date of the appearance of the caterpillar is in May.

It feeds on the lime, apple, rose, &c.

CIDARIA MIARIA.

GREEN CARPET.

Plate XXXIV. Figure 4.

Localities for this species are York, Brighton, Eastham near Birkenhead, Scarborough, Falmouth, Barnstaple, Bristol, Stowmarket, Black Park, Cambridge, Darlington, Tenterden, Worcester, Edinburgh, Exeter, Halton, Carron, Worthing, Huddersfield, Kingsbury, Dunoon, Glasgow, Manchester, Lewes, Newcastle-on-Tyne.

The perfect insect appears in September and October.

The caterpillar is green, with two projecting points from the hind segments.

The date of the appearance of the caterpillar is in August.

It feeds on the alder, oak, and birch.

CIDARIA PICARIA.

CLOAKED CARPET.

Plate XXXIV. *Figure 5.*

Localities for this species are Brighton, Barnstaple, Exeter, Poynings, Pembury, Isle of Wight, Stowmarket, Lewes, Malvern, Tenterden, Darlington, Conway.

The situations where it is found are woods.

The perfect insect appears in June and July.

CIDARIA CORYLARIA.

BROKEN-BARRED CARPET.

Plate XXXIV. *Figure 6.*

Localities for this common species are York, Brighton, Norbury Park, Arundel, Glasgow, Rannock, Pitochrie.

The situations where it is found are woods.

The perfect insect appears in June.

The caterpillar is of a rose colour, the back tinged with yellowish, with two small points on the hind segment.

The date of the appearance of the caterpillar is September.

It feeds on the lime and the sloe.

CIDARIA SAGITTARIA.

Plate XXXIV. *Figure* 7.

Localities for this species are Peterborough, and the fens generally of Cambridgeshire and Huntingdonshire.

The situations where it is found are the fens.

The perfect insect appears in July.

I have to thank Mr. Joseph Steele, of Congleton, for assistance to me in this work.

CIDARIA RUSSARIA.

MOTTLED CARPET. COMMON MOTTLED CARPET.

Plate XXXIV. *Figure* 8.

Localities for this very common species are York, Brighton, Anstey, Faversham, Exeter, Isle of Man, Arran, Dunoon.

The situations where it is found are hedge sides, gardens, woods, etc.

The perfect insect appears in May and August.

The caterpillar is yellowish green, with a darker line along the back, and the side line sometimes red.

The date of the appearance of the caterpillar is in April and August.

It feeds on various plants.

I have also to thank Mr. Edward Brown, of York, for many obliging assistances.

CIDARIA IMMANARIA.

Plate XXXIV. *Figure* 9.

Localities for this very common species are York, Barnstaple, Isle of Man.

The situations where it is found are wooded places, woods, gardens, lanes, and hedge sides.

The perfect insect appears in July and September.

CIDARIA SUFFUMARIA.

WATER-CARPET.   SMALL WATER-CARPET.

Plate XXXIV. *Figure* 10.

Localities for this generally distributed species are York, Brighton, Falmouth, Barnstaple, Bristol, Darlington.

The situations where it is found are woods, gardens, and hedge sides.

The perfect insect appears in May.

The caterpillar is green, paler on the back.

The date of the appearance of the caterpillar is in April.

COREMIA RETICULARIA.

Plate XXXIV. *Figure* 11.

Localities for this species are the Lake Districts.

The perfect insect appears in August.

CIDARIA SILACEARIA.

SMALL PHŒNIX.

Plate XXXIV. *Figure* 12.

Localities for this species are York, Scarborough, Hale near Liverpool, Brighton, Pembury, Epping, Bristol, Darlington, Stowmarket, Arundel, Exeter, Worcester, Halton, Newcastle-on-Tyne, Lewes, Newnham, Preston, Ambleside, Dunoon, Renfrew, Edinburgh.

The situations where it is found are woods, lanes, and hedge sides.

The perfect insect appears in May and June.

The caterpillar is pale green, with a white line on each side below the back edged above with elongated red spots; the side line yellowish-white.

The date of the appearance of the caterpillar is in September.

It feeds on the aspen.

CIDARIA PRUNARIA.

CLOUDED CARPET.   LARGE WATER-CARPET.

Plate XXXIV. *Figure* 13.

Localities for this rather common species are York, Brighton, Faversham, Barnstaple, Worcester, Sudbury.

The situations where it is found are lanes and gardens.

The perfect insect appears in July and August.

The caterpillar is green, with distinct white spots; the third segment with a reddish band spotted with white; the fifth, sixth, seventh, eighth, ninth, and tenth segments each with a white spot edged with red on the back.

The date of the appearance of the caterpillar is in May.

It feeds on the gooseberry tree and the currant tree.

CIDARIA TESTARIA.

Plate XXXIV. *Figure* 14.

Localities for this very common species are York, Brighton, Manchester, Dunoon, Hammersmith, Lewisham, Faversham, Barnstaple, Ilfracombe.

The situations where it is found are woods and wooded lanes.

The perfect insect appears in July, August, and September.

The caterpillar is pale yellowish white, with a dark line along the back; the side line whitish, edged on its upper part with grey.

The date of the appearance of the caterpillar is June. It feeds on the aspen.

CIDARIA POPULARIA.

THE POPLAR CARPET.

Plate XXXIV. *Figure* 15.

Localities for this species are York, Huddersfield, Darlington, Isle of Man, Brighton, Pembury, Storeton near Birkenhead, Newcastle-on-Tyne, Manchester, Faversham, Tenterden, Barnstaple, Lynton, West Looe, Stowmarket, Edinburgh, Glasgow, Arran, Inverness, Waterford.

The situations where it is found are woods and wooded lanes.

The perfect insect appears in July and August.

The caterpillar is pale green, with a reddish line along the back, broadest at the inner edge of each segment.

The date of the appearance of the caterpillar is in May. It feeds on the sallow and the bilberry.

CIDARIA FULVARIA.

BARRED YELLOW.

Plate XXXV. *Figure* 1.

Localities for this beautiful and sufficiently common species are York, Nunburnholme, Charmouth, Brighton, Humberstone, Falmouth, Faversham, Birkenhead, Dorking, West Looe.

The situations where it is found are gardens, lanes, and hedge sides.

The perfect insect appears in July and August.

The caterpillar is whitish green on the back, and dark green on the sides; the side line whitish, as are the interstices between the segments.

The date of the appearance of the caterpillar is in May. It feeds on the rose, as well so beautiful a species may.

CIDARIA PYRALIARIA.

BARRED STRAW.

Plate XXXV. *Figure* 2.

Localities for this also common species are York, Brighton, Humberstone, Faversham, Isle of Wight Barnstaple, Lynton.

The situations where it is found are gardens, lanes, hedge sides, &c.

The perfect insect appears in July and August.

The caterpillar is yellowish green, with a paler line along the back, as also are the interstices between the segments.

The date of the appearance of the caterpillar is in May It feeds on the whitethorn.

CIDARIA DOTARIA.

THE SPINACH.

Plate XXXV. *Figure* 3.

Localities for this species are York, Brighton, Pembury, Birkenhead, Falmouth, Exeter, Stowmarket, Newnham, Sudbury, Cambridge, Manchester, Stowmarket, Preston, Newcastle-on-Tyne, Paisley.

The situations where it is found are gardens.

The perfect insect appears in June and July.

The caterpillar is green, with the side line of a paler shade of the same.

The date of the appearance of the caterpillar is in April and May.

It feeds on the currant tree.

PELURGA COMITARIA.

DARK SPINACH.

Plate XXXV. *Figure* 4.

Localities for this species are York, Scarborough, Brighton, Monkton, Newcastle-on-Tyne, Ipswich, Cambridge, Edinburgh, Kinsbury, Birkenhead, Darlington, Manchester, Pembury.

The perfect insect appears in July.

The caterpillar is greenish grey, speckled with black, the side line yellowish grey and broad.

The date of the appearance of the caterpillar is in September and October.

It feeds on the goose-foot (*Chenopodium polyspermum.*)

The chrysalis is found under the earth.

EUBOLIA CERVINARIA.

THE MALLOW MOTH.

Plate XXXV. *Figure* 5.

Localities for this neat species are York, Huddersfield, Rotherham, Stowmarket, Malton, Bromsgrove, Brighton, Falmouth, Dulwich, Newcastle-upon-Tyne, Cambridge, Lower Guiting, Lewisham, Newnham, Darlington, Plumstead, Barnstaple, Bristol, Exeter, Black Park, Birkenhead, Worcester, Preston, Ardrossan, Edinburgh.

The situations where it is found are gardens.

The perfect insect appears in September and October.

The caterpillar is green; the head, legs, and interstices between the segments, whitish green.

The date of the appearance of the caterpillar is in June and July.

It feeds on the mallow and the hollyhock.

The chrysalis is found below the earth.

EUBOLIA MENSURARIA.

SMALL MALLOW.

Plate XXXV. *Figure* 6.

Localities for this plentiful species are York, Brighton, Anstey, Falmouth, Worcester, Birkenhead, Isle of Man, Golspie.

The situations where it is found are commons and waste grassy places, &c.

The perfect insect appears in June, July, and August.

The caterpillar is described as yellowish green.
The date of the appearance of the caterpillar is in June.
It feeds on the grass.
The chrysalis is found under the ground.

EUBOLIA PALUMBARIA.

THE BELLE MOTH.

Plate XXXV. *Figure* 7.

Localities for this species are York, Stockton Common near York, Huddersfield, Bromsgrove Lickey, Falmouth, Edinburgh, Cambridge, Faversham, Charmouth, Barnstaple, Lyndhurst, Manchester, Sudbury, Birkenhead, Lewes, Halton, Newcastle-on-Tyne, Isle of Man, Glasgow, Exeter, Darlington, Newnham, Scarborough.

The situations where it is found are commons, heaths, cliffs, and waste grassy places.

The perfect insect appears in May, June, July, and August.

The caterpillar is whitish grey, with three lines of dark grey on each side.

The date of the appearance of the caterpillar is in March and April.

It feeds on the heath *(Calluna vulgaris,)* and clover.

The chrysalis is subterranean.

EUBOLIA BIPUNCTARIA.

CHALK CARPET.   CHALK MOTH.

Plate XXXV.   *Figure* 8.

Localities for this species are York, Scarborough, Castle Eden Dene, Charmouth, Pinhay Cliff, Hoylake near Birkenhead, Brighton, Newcastle-on-Tyne, Bristol, Cambridge, Faversham, Halton, Lewes, Dover, West Looe.

The situations where it is found are chalky cliffs, chalk pits, and waste grassy places.

The perfect insect appears in July and August.

The caterpillar is pale brownish grey, with a darker line along the back, and another below it on each side.

The date of the appearance of the caterpillar is in June and July.

It feeds on clover, &c.

The chrysalis is found below the surface of the earth.

EUBOLIA LINEOLARIA.

THE OBLIQUE STRIPED.

Plate XXXV.   *Figure* 9.

Localities for this species are Brighton, New Brighton, Cambridge, Deal, Lewes, Littlehampton in Sussex, Birkenhead.

The situations where it is found are heaths.

The perfect insect appears in May, June, July, and August.

The caterpillar is pale yellowish brown, with a dark brown line along the back; the side line also dark brown, edged beneath with pale yellow.

The date of the appearance of the caterpillar is in May, and also in September.

It feeds on the bedstraw *(Galium verum)*.

The chrysalis is found beneath the earth.

CARSIA IMBUTARIA.

MANCHESTER TREBLE-BAR.

Plate XXXV. *Figure* 10.

Localities for this species are Simonswood Moss and Bickerstaffe Moss near Liverpool, Chat Moss near Manchester, Newcastle-on-Tyne, Preston, Inverness, Dunoon.

The situations where it is found are heathy places.

The perfect insect appears in July.

The caterpillar is reddish yellow, with three stripes of a violet colour along the back; the side line yellowish white.

The date of the appearance of the caterpillar is in June.

It feeds on the cranberry *(Vaccinium oxycoccos)*.

The chrysalis is found in a slight cocoon among moss.

ANAITIS PLAGIARIA.

TREBLE BAR. SLENDER TREBLE BAR.

Plate XXXV. *Figure* 11.

Localities for this species are Scarborough, Brighton, Huddersfield, Anstey, Faversham, Milstead, Stowmarket, Childwall near Liverpool, Tenterden, Weston-super-Mare, Barnstaple, Newnham, Halton, Kingsbury, Manchester, Sudbury, Bristol, Newcastle-on-Tyne, Exeter, Wisbeach, Cambridge, Edinburgh, Gourock, Lyndhurst, Ardrossan, Lewes, Arran, Bute.

The situations where it is found are woods and grassy wastes.

The perfect insect appears in June, July, August, and even to September.

The caterpillar is reddish brown, with a slender black interrupted line along the back; the side line bright yellow.

The date of the appearance of the caterpillar is in July.

It feeds on the St. John's wort *(Hypericum calycinum)*.

LITHOSTEGE NIVEARIA.

Plate XXXV. *Figure* 12.

Localities for this species are Thetford and Brandon in Suffolk.

The perfect insect appears in June and July.

CHESIAS SPARTIARIA.

THE STREAK.

Plate XXXV. *Figure* 13.

Localities for this species are York, Hooton near Birkenhead, Brighton, Wavendon, Leatherhead, Darlington, Newnham, Coombe in Surrey, Newcastle-on-Tyne, Carlisle, Edinburgh, Stowmarket, Glasgow.

The situations where it is found are places where the wild broom grows.

The perfect insect appears in September and October.

The caterpillar is dark green, with a darker line edged with pale green along the back; the side line white.

The date of the appearance of the caterpillar is in June.

It feeds on the broom.

CHESIAS OBLIQUARIA.

THE CHEVRON.

Plate XXXV. *Figure* 14.

Localities for this species are Brighton, Newcastle-on-Tyne, Dartford, Stowmarket, Coombe Warren in Surrey, Weybridge, Sudbury.

The situations where it is found are near woods where broom grows.

The perfect insect appears in May, also in August?

The caterpillar is green, the back darker, the hind segment with two points on it.

The date of the appearance of the caterpillar is in August.

It feeds on the broom.

TANAGRA CHŒROPHYLLARIA.

THE CHIMNEY-SWEEP.   CHIMNEY-SWEEPER.   SWEEP.

Plate XXXV. *Figure* 15.

Localities for this species are York, Scarborough, Nunburnholme, near Warter Priory, Howsham, Huddersfield, Brighton, Darlington, Edinburgh, Kingsbury, Lyndhurst, Bromsgrove, Barnstaple, Lower Guiting, Newcastle-on-Tyne, Manchester, Plymouth, Lynton, Cambridge, Glasgow, Pembury, Newnham.

The situations where it is found are grass meadows, and open places in and near woods, etc.

The perfect insect appears in June and July.

The caterpillar is velvet green.

The date of the appearance of the caterpillar is in May, and also in July.

It feeds on the chervil, (*Chærophyllum temulentum.*)

The chrysalis is placed in a slight cocoon.

DREPANULÆ.

PLATYPTERYX LACERTULA.

Plate XXXVI. *Figure* 1.

Localities for this species are Langwith and Askham Bog near York, Scarborough, Blean Woods near Canterbury, Brighton, Horndean, Stowmarket, Worthing, Bristol, Worcester, Exeter, Lower Guiting, Shrewsbury, Crewe, Carlisle, Halton, Epping, Lewes, Lyndhurst, Chat Moss, Manchester, Tenterden, Birch Wood and Simonswood Moss near Liverpool, Wavendon, Castle Eden Denc.

The situations where it is found are birch plantations.

The perfect insect appears in March, April, May, June, July, and August.

The caterpillar is pale brown and yellowish, with spots and clouds of darker brown, the third, and fourth, and twelfth segments with two protuberances.

The date of the appearance of the caterpillar is in June and September.

It feeds on the birch.

DREPANULA SICULA.

Plate XXXVI. *Figure* 2.

Localities for this species are Brighton, and Leigh Wood near Bristol.

The perfect insect appears at the end of May, and in June.

The caterpillar is reddish-brown, with a broad pale yellow stripe spotted with brown along the back, the spots most abundant on the second, third, and fourth segments, the last named having also two protuberances on it.

The date of the appearance of the caterpillar is in May and June.

It feeds on the oak, the lime, and the birch.

DREPANULA FALCULA.

'PEBBLE HOOK-TIP.

Plate XXXVI. *Figure* 3.

Localities for this species are Langwith and Stockton Forest near York, Brighton, Blean Woods near Canterbury, Buttercrambe Moor near Stamford-Bridge, Worthing, Bysing Wood near Faversham, Carlisle, Bristol, Linwood near Market Rasen, Halton, Tenterden, Isle of Wight, Lower Guiting, Blandford, Epping, Huddersfield, Worcester, Lewes, Preston, Shrewsbury, Newnham, Lyndhurst, Wisbeach, Stowmarket, Kilmun, Chat Moss near Manchester, Scarborough, Birch Wood and Simonswood Moss near Liverpool.

The situations where it is found are woods and wooded parts of commons.

The perfect insect appears in May, June, August, and September?

The caterpillar is pale green, with a stripe of dark reddish brown on the back; the second, third, fourth, fifth, and sixth segments each with two small protuberances.

The date of the appearance of the caterpillar is in May and September.

It feeds on the birch, the alder, the aspen, the willow, and the oak.

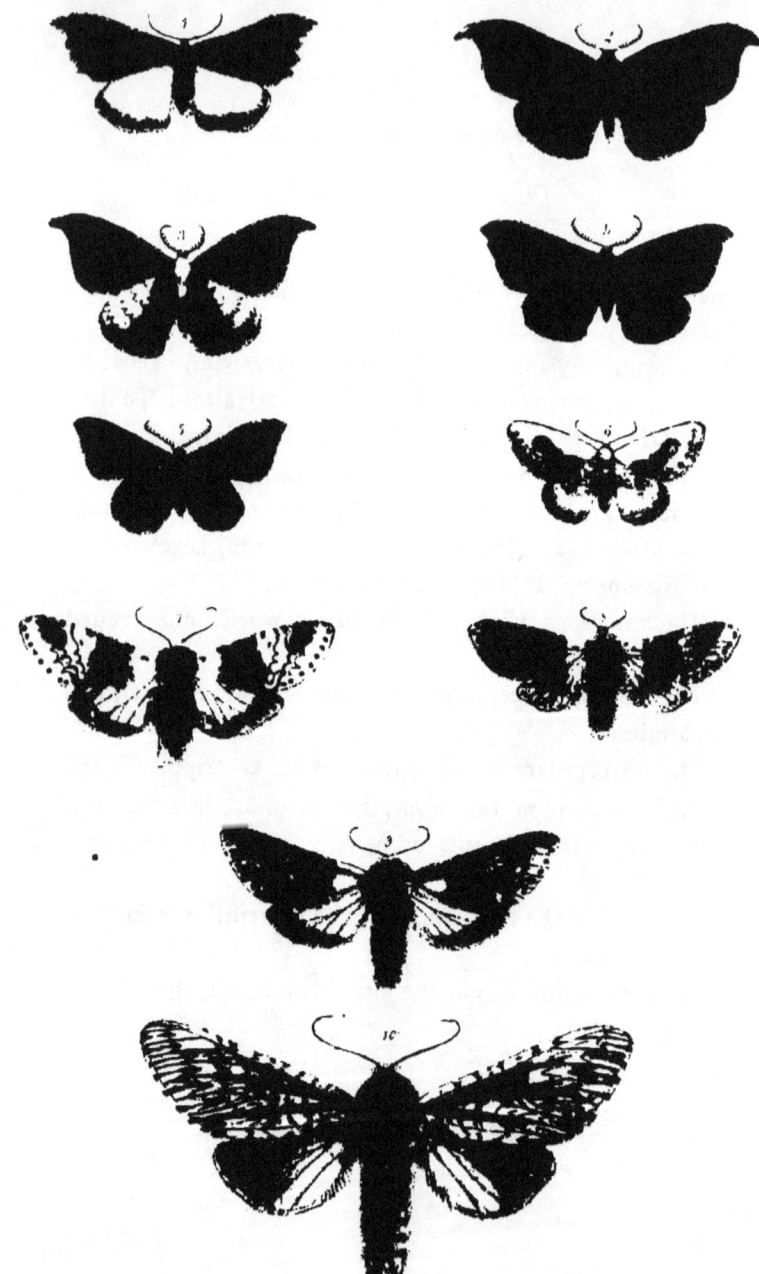

DREPANULA HAMULA.

OAK HOOK-TIP.

Plate XXXVI. *Figure* 4.

Localities for this species are Brighton, Epping, Carlisle, Lewisham, Bristol, Dulwich, Lewes, Newnham, Stowmarket, Worcester, Lympstone.

The situations where it is found are oak and birch woods.

The perfect insect appears in May, June, July, and August. August 28th.

The caterpillar is greyish, with a broad stripe along the back, of a greenish brown colour on the second, third, fourth, twelfth, and thirteenth segments, and yellowish brown on the other, edged with yellow on each side; the fourth segment with two prominences on it.

The date of the appearance of the caterpillar is in June and September.

It feeds on the oak and the birch.

This species is attracted by light.

DREPANULA UNGUICULA.

BARRED HOOK-TIP.

Plate XXXVI. *Figure* 5.

Localities for this species are Richmond Park, Epping Forest, the New Forest, Brighton, Stowmarket, Mickleham, Tenterden, Sevenoaks, Wavendon, Newnham, Bristol, Dursley, Halton, Worcester.

The perfect insect appears in May and June; also in August.

The caterpillar is reddish brown, with a dark brown stripe along the back on the sixth, seventh, eighth, ninth, and tenth segments; the second, third, and fourth segments with a yellow line on each side, meeting on the back of the fifth, the fourth segment having a raised prominence on it.

The date of the appearance of the caterpillar is in June and September.

It feeds on the beech.

CILIX SPINULA.

CHINESE CHARACTER. GOOSE-EGG.

Plate XXXVI. *Figure* 6.

Localities for this species are Nunburnholme, York, Sutton-on-Derwent, Bromsgrove, Worcester, Humberstone, Brighton, Falmouth, Bisterne, Dorking.

The situations where it is found are hedge sides and woods.

The perfect insect appears in May, June, and August.

The caterpillar is dull chocolate brown, with prominences on the third and fourth segments.

The date of the appearance of the caterpillar is at the end of May, in June, and on to the end of July.

It feeds on the blackthorn and the whitethorn.

DICRANURIDÆ.

CERURA BICUSPIS.

Plate XXXVI. *Figure* 7.

Localities for this species are Stockton Forest, York, Ripon, Scarborough, Goathland near Whitby, Dublin, Preston, Middlesborough.

The perfect insect appears in May, and the beginning of June to the end. May 10th, 27th, June 20th.

The caterpillar is yellowish green, with a red stripe along the back, narrowed to the fourth segment, and widened again to the eighth, diminishing then on to the tail.

The date of the appearance of the caterpillar is in August.

It feeds on the beech, the alder, and the birch.

The chrysalis is found in a hard cocoon attached to the bark of trees.

CERURA FURCULA.

KITTEN MOTH.

Plate XXXVI. *Figure* 8.

Localities for this species are Buttercrambe Moor near York, Lewes, Lewisham; Norwood, Ashford in Kent, Taunton, Ilfracombe, Worcester, Durham, Cambridge, Lyndhurst, Carlisle, Bristol, Glasgow, Manchester, Stowmarket, Worthing, Lower Guiting, Darlington, Epping, Edinburgh, Preston, Halton, Plymouth.

The perfect insect appears about the end of June and the beginning of August.

The caterpillar is yellowish green, dotted with darker green and reddish, and having a reddish stripe along the back spotted and edged with yellow, the red interrupted on the sixth and seventh segments.

The date of the appearance of the caterpillar is in August.

It feeds on the sallow and the willow.

The chrysalis is found enclosed in a very hard cocoon attached to the bark of the tree whose leaves had formed the food of the larva.

CERURA BIFIDA.

Plate XXXVI. *Figure 9.*

Localities for this species are York, Huddersfield, Rotherham, Sutton-on-Derwent, Stockton Forest, Askham Bog and Langwith near York, Brighton, Lewisham, Ashford, Halton, Kingsbury, Manchester, Horndean, Taunton, Sherwood Forest, Carlisle, Preston, Stowmarket, Lower Guiting, Cambridge, Epping, Birkenhead, Bristol, Burton-on-Trent, Wavendon, Hammersmith, and near London.

The perfect insect appears in June, July, and the beginning of August.

The caterpillar is pale green dotted with brown, with a broad stripe along the back, on the second and third segments, on the latter of which it ends in a point, widening again to the eighth, and then diminishing to the thirteenth, where it finally again widens.

The date of the appearance of the caterpillar is in August and September.

It feeds on the aspen and the poplar.

The chrysalis is found in a very hard cocoon, fixed on the bark of the tree on which the caterpillar feeds.

CERURA VINULA.

PUSS MOTH.

Plate XXXVI. *Figure* 10.

Localities for this extensively distributed species are York, Nafferton, Sutton-on-Derwent, Stockton Forest, Bromsgrove, Brighton, Queenstown, Falmouth, Canterbury, Faversham, Ashford, Cambridge, Bristol, Carlisle, Exeter, Lynton in Devonshire, Sherwood Forest.

The situations where it is found are gardens, pool sides, and other places where the willow and the poplar grow.

The perfect insect appears in May and June.

The caterpillar is dark green, with a prominence on the fourth segment, then a brown belt bordered with white along the back, widest at the eighth segment, and then narrowing gradually to the double tail.

The date of the appearance of the caterpillar is in July and August.

It feeds on the willow, the sallow, and the poplar of various sorts.

The chrysalis is found enclosed in a very hard cocoon attached to the bark of the tree whose leaves have afforded nourishment to the larva.

STAUROPUS FAGI.

LOBSTER MOTH.

Plate XXXVII. *Figure* 1.

Localities for this rather rare species, which derives its name from a supposed resemblance of the caterpillar to the shell-fish so called, are Worcester, Hammersmith, Lewes, Newnham, West Wickham, West Looe, Sherwood Forest, Blandford, Black Park, Epping, Sevenoaks, Cowes, Exeter, Lyndhurst, Arundel, Halton, Plymouth, Henley-on-Thames, Dursley, Bristol.

The situations where it is found are woods.

The perfect insect appears in June and July, the middle of the former month and the beginning of the latter.

The caterpillar is reddish brown, with two protuberances on the fourth, fifth, sixth, seventh, and ninth segments; the two hind segments carried erect, and each bearing a short tail.

The date of the appearance of the caterpillar is in August and September.

It feeds on the beech, the oak, and the birch.

PETASIA CASSINEA.

THE SPRAWLER.

Plate XXXVII. *Figure* 2.

Localities for this species are York, Doncaster, Carlisle, Tenterden, Brighton, Rannock, Lewisham, Black Park, Sidmouth, Darlington, Epping, Marlow, Dumfries, Exeter,

Plate XXXVII

Lyndhurst, Bisterne, Halton, Plymouth, Preston, Worcester, Shrewsbury, Lower Guiting, Cambridge, Stowmarket, Bristol, Burton-on-Trent, Peterborough.

The situations where it is found are woods.

The perfect insect appears in October.

The caterpillar is yellowish-green, with three white lines along the back, and a yellow one on each side, meeting at the end..

The date of the appearance of the caterpillar is in May and June; also in August.

It feeds on the birch, the elm, the lime, the sallow, and the oak.

PETASIA NUBECULOSA.

Plate XXXVII. *Figure* 3.

Localities for this species are Rannock in Perthshire.

The perfect insect appears in April.

The caterpillar is green, with raised white dots.

The date of the appearance of the caterpillar is in May and June.

It feeds on the birch (and the elm.)

PSEUDO-BOMBYCES.—PYGÆRIDÆ.

PYGÆRA BUCEPHALA.

BUFF TIP.

Plate XXXVII. *Figure* 4.

Localities for this plentiful species are York, Driffield, Nafferton, Pocklington, Sutton-on-Derwent, Bromsgrove,

Queenstown, Brighton, Anstey, Falmouth, Bisterne, Manchester, Sudbury, Durham, Isle of Man, Dunoon.

The situations where it is found are by hedgerow timber, and elsewhere, especially where the elm tree grows.

The perfect insect appears in June and July.

The caterpillar is rather dark greenish yellow, with a broad blackish line along the back, and three others on each side.

The date of the appearance of the caterpillar is in August and September.

It feeds on the elm principally, but also on the oak and the birch, the sallow, the lime, and the nut-tree.

The chrysalis is found beneath the surface.

CLOSTERA CURTULA.

CHOCOLATE TIP.

Plate XXXVII. *Figure 5.*

Localities for this species are York, Stockton Forest, Buttercrambe Moor, Carlisle, Canterbury, Bristol, Forest Hill, Darenth Wood, Eltham, Isle of Wight, Newmarket, Burton-on-Trent, Epping, Wicking Fen, Halton, Worcester, and near London.

The situations where it is found are woods.

The perfect insect appears in May and July—July 20.

The caterpillar is dull reddish black, powdered with white, with two rows of orange-coloured raised spots on each side, and the fifth and twelfth segments each with a velvet black prominence on it.

The date of the appearance of the caterpillar is in June and July, August and September—July 2.

It feeds on the aspen, the sallow, and the poplar.

CLOSTERA ANACHORETA.

Plate XXXVII. *Figure* 6.

Localities for this species are near London.

The caterpillar is yellowish, or reddish grey, with longitudinal interrupted lines, and alternating black and yellow spots on the sides, a small reddish prominence on the twelfth segment, and a larger one on the fifth, also a white spot on each side. This is one account; another, and I believe a more correct one, is as follows:—It is slightly covered with yellowish hair and has a broad pale buff band along the back, below which on either side is a slate-black band, with some dull orange spots, and beneath this the prevailing colour is orange, the spots above being brighter, the head black, and on the fourth segment is a dull pink-red raised spot surrounded by a black patch, in which, on either side of the prominence, is a conspicuous white spot; on the lesser segment is another similar but smaller prominence.

The date of the appearance of the caterpillar is from June to October.

It feeds on the willow and the poplar.

CLOSTERA RECLUSA.

SMALL CHOCOLATE TIP.

Plate XXXVIII. *Figure* 1.

Localities for this species are York, Canterbury, Cambridge, Dublin, the Isle of Arran, Brighton, Black Park, Howden, Bristol, Peterborough, Epping, Glasgow, Halton, Pembury, Edinburgh, Ben Nevis, Carlisle, Horndean.

The perfect insect appears in May.

The caterpillar is blackish grey; the back greenish grey, white on the sides, a row of yellow spots and a double yellow line, and a black raised spot on the fifth and twelfth segments.

The date of the appearance of the caterpillar is in September.

It feeds on the sallow.

NOTODONTIDÆ.

GLUPHISIA CRENATA.

Plate XXXVIII. *Figure* 2.

Localities for this very rare species are Ongar Park Wood, Halton in Buckinghamshire, and at Epping.

The situations where it is found are woods.

The perfect insect appears in May and June.

The caterpillar is pale green, with a line along the back, spotted with rust colour, and bordered on each side by a yellow line.

The date of the appearance of the caterpillar is in August.

It feeds on the poplar.

PTILOPHORA PLUMIGERA.

Plate XXXVIII. *Figure* 3.

Localities for this species are Darenth Wood, Birch Wood, Chatham, Henley-on-Thames, Marlow, and Halton.

The situations where it is found are woods.

The caterpillar is pale green, with a line of bluish green

along the back, bordered on each side by another of white, and two very narrow whitish lines above the legs, which are pale green, as is the head.

The date of the appearance of the caterpillar is in May and June.

It feeds on the maple.

PTILODONTIS PALPINA.

PALE PROMINENT.

Plate XXXVIII. *Figure 4.*

Localities for this species are York, Sutton-on-Derwent, Brighton, Lewisham, Shooter's Hill, Norwood, Wavendon, Henfield, Tenterden, Lewes, Whippingham, Shrewsbury, Worcester, Stowmarket, Wisbeach, Plymouth, Ipswich, Newark, Derby, Carlisle, Lyndhurst.

The situations where it is found are woods, gardens, &c.

The perfect insect appears in May, June, July, and September.

The caterpillar is pale green, with four interrupted white lines along the back, and a yellow line on the sides, edged on its upper part, with black on the second and fourth segments.

The date of the appearance of the caterpillar is in June and October.

It feeds on the aspen, the sallow, and the poplar.

NOTODONTA CAMELINA.

COXCOMB PROMINENT.

Plate XXXVIII. *Figure* 5.

Localities for this species are York, Sheffield, Sutton-on-Derwent, Longwith, Askham Bog, Sherwood Forest, Dunham Park, Shooter's Hill, Durham, Black Park, Glasgow, Canterbury, Maidstone, Faversham, Henfield, Bisterne, Horndean, Isle of Wight, Malvern.

The perfect insect appears in June, July, and August.

The caterpillar is greenish, with a yellowish green line along the sides, the twelfth segment with two small prominences tipped with red.

The date of the caterpillar is in August, September, and October.

It feeds on the birch, the hazel, &c.

NOTODONTA CUCULLINA.

MAPLE PROMINENT.

Plate XXXVIII. *Figure* 6.

Localities for this species are Arundel, Marlow, Henley, Tring, Lynton, Halton.

The perfect insect appears in May.

The caterpillar is green, or pale reddish, with a broad mark of a darker colour extending backwards to the fifth segment, the twelfth segment with a red tip.

The date of the appearance of the caterpillar is in August and September.

It feeds on the maple.

NOTODONTA CARMELITA.

Plate XXXVIII. *Figure* 7.

Localities for this species are Ongar Park Wood, Darenth Wood, Black Park, Keswick, Birch Wood, West Wickham Wood, Epping, Carlisle.

The situations where it is found are woods.

The perfect insect appears at the end of March, and in April and May—May 4, 10, 15.

The caterpillar is green, dotted with numerous raised yellow spots, the side line yellowish white.

The date of the appearance of the caterpillar is in June. It feeds on the birch.

NOTODONTA BICOLOR.

Plate XXXVIII. *Figure* 8.

Localities for this species are Killarney.

The situations where it is found are birch woods.

The perfect insect appears in May and June, to the end of the latter month and in July.

The caterpillar feeds on the birch.

NOTODONTA DICTÆA.

SWALLOW PROMINENT.

Plate XXXIX. *Figure* 1.

Localities for this species are York, Rotherham, Langwith, Huddersfield, Ramsgate, Dover, Lewes, Bisterne Horndean, Darlington, Edinburgh, Worcester, Durham

Carlisle, Burton-on-Trent, Ipswich, Witney, Birkenhead, Cambridge, Blandford, Lower Guiting, Epping, Halton, Glasgow, Derby, Bristol, Manchester, Plymouth, Preston, Carron, Stirling, Scarborough, Sheffield, Teignmouth, Tenterden, Shooter's Hill near London.

The perfect insect appears in May, June, and July.

The caterpillar is greenish white, with a yellow stripe on each side, otherwise dull brown.

The date of the appearance of the caterpillar is in August.

It feeds on the birch, the poplar, and the willow.

NOTODONTA DICTÆOIDES.

SWALLOW-LIKENESS PROMINENT.

Plate XXXIX. *Figure* 2.

Localities for this species are York, Rotherham, Langwith, Brighton, Shooter's Hill, Wandsworth, West Wickham, Cockermouth, Bristol, Halton, Horndean, Birkenhead, Lyndhurst, Malvern, Edinburgh, Manchester, Preston, Derby, Epping, Plymouth, Perth.

The perfect insect appears in May and June.

The caterpillar is deep glossy brown, showing a tinge of purple, and with a broad yellow band on each side.

The date of the appearance of the caterpillar is in August and September.

It feeds on the birch.

NOTODONTA DROMEDARIUS.

IRON PROMINENT.

Plate XXXIX. Figure 3.

Localities for this species are York, Langwith, Askham Bog, Shooter's Hill in Kent, Norwood in Surrey, Birkenhead, Bristol, Edinburgh, Scarborough, Carlisle, Black Park, Brighton, Lower Guiting, Lyndhurst, Horndean, Epping, Glasgow, Manchester, Stowmarket, Halton, Plymouth, Wavendon, Derby, Sheffield, Worcester, Preston, and near London.

The situations where it is found are woods.

The perfect insect appears in June.

The caterpillar is yellowish green, with a purple brown stripe along the back, on the second, third, and fourth segments, the fifth, sixth, seventh, eighth, and twelfth with small prominences.

The date of the appearance of the caterpillar is in August. It feeds on the birch.

NOTODONTA TRITOPHUS.

DARK PROMINENT.

Plate XXXIX. Figure 4.

Localities for this species are near Dublin.

The perfect insect appears in May, August. August 10.

The caterpillar is dark green, with prominences on the fifth, sixth, seventh, and twelfth segments, a reddish streak along the back, as far as the fifth segment, and an interrupted reddish streak on the sides.

The date of the appearance of the caterpillar is in July and September.

It feeds on the aspen, the poplar, and the birch.

NOTODONTA ZICZAC.

PEBBLE PROMINENT.

Plate XXXIX. *Figure* 5.

Localities for this species are York, Scarborough, Brighton, Bromsgrove, Worcester, Norwood, Carlisle, Worthing, Wavendon, Tenterden, Arundel, Preston, Manchester, Stowmarket, Lyndhurst, Lewes, Huddersfield, Horndean, Halton, Glasgow, Exeter, Cambridge, Birkenhead, Bristol, Bisterne, Epping, Darlington, Newark, Sherwood Forest, Bolton in Lancashire, Fort-William, Inverness, Burton-on-Trent.

The perfect insect appears in May, June, and August.

The caterpillar is variously grey, violet grey, or reddish brown, the three last segments reddish brown, the sixth, seventh, and eighth with an angular prominence, and on the sides are three pale stripes.

The date of the appearance of the caterpillar is in June, September, and October.

It feeds on the poplar and the sallow.

NOTODONTA TREPIDA.

GREAT PROMINENT.

Plate XXXIX. *Figure* 6.

Localities for this fine species are York, Doncaster, Stockton Forest, Brighton, West Wickham Wood, Cockermouth, Darenth Wood, Taunton, Carlisle, Tenterden, Shooter's Hill, Ipswich, Sherwood Forest, Black Park, Shrewsbury, Manchester, Epping, Bisterne, Lyndhurst, Horndean, Halton, Exeter, Plymouth, and near London.

The situations where it is found are woods.

The perfect insect appears in April and May—April 27; May 6-8.

The caterpillar is yellowish green, with two white lines on the back, and on the side of each segment an oblique red stripe, margined with yellow.

The date of the appearance of the caterpillar is in July, August, and September.

It feeds on the oak.

NOTODONTA CHAONIA.

LUNAR MARBLED BROWN.

Plate XXXIX. *Figure* 7.

Localities for this species are York, Brighton, Sherwood Forest, Wavendon, Eastham and Hooton near Birkenhead, Black Park, Tring, Bristol, Epping, Halton, Wymondham, Lyndhurst, Manchester, Plymouth, Worthing, Cockermouth, Worcester, Torwood near Stirling, Carlisle, and Shooter's Hill, near London.

The situations where it is found are woods.

The perfect insect appears in May.

The caterpillar is whitish green, with a yellow line on each side of the back, and a broader one on the sides.

The date of the appearance of the caterpillar is in July.

It feeds on the oak.

NOTODONTA DODONEA.

MARBLED BROWN.

Plate XXXIX. *Figure* 8.

Localities for this species are Brighton, Taunton, Sherwood Forest, Ipswich, Lower Guiting, Halton, Worcester, Lyndhurst, Delamere Forest, Cockermouth, and near London.

The situations where it is found are woods.

The perfect insect appears in May and June. May 19.

The caterpillar is pale bluish green, with two white lines on the back, a row of white dots on the side, and a yellowish side line.

The date of the appearance of the caterpillar is in July and August.

It feeds on the oak and the birch.

DILOBA CŒRULEOCEPHALA.

FIGURE-OF-EIGHT MOTH.

Plate XXXIX. *Figure* 9.

Localities for this common species are York, Ripon, Nunburnholme, Sutton-on-Derwent, Bromsgrove, Perth, Lewisham, Brighton, Bisterne, Carlisle, Dumfries.

The situations where it is found are hedge-sides, &c.

The perfect insect appears in September.

The caterpillar is pale yellow, with a pale green, or pale blue band on the sides, the head blue, spotted with black.

The date of the appearance of the caterpillar is in June.

It feeds on the black-thorn, wild plum, or sloe.
The chrysalis is enclosed in a web attached to a stem.

TRIFIDÆ.

BOMBYCIFORMES.

THYATIRA DERASA.

BUFF ARCHES.

Plate XL. *Figure* 1.

Localities for this species are York, Doncaster, Sutton-on-Derwent, Charmouth, Bristol, Faversham, Brighton, Canterbury, Dartford, Dorking, Clapham, Peterborough, Rotherham, Black Park, Arundel, Barnstaple, Gloucester, Monks Wood, Llanelly, Durham, Killarney, Birkenhead, Burton-on-Trent, Cambridge, Lower Guiting, Kingsbury, Huddersfield, Exeter, Worthing, Lyndhurst, Worcester, Tenterden, Scarborough, Shrewsbury, Stowmarket, Manchester, Plymouth.

The perfect insect appears in June and July.

The caterpillar is fine dark brown, with a conspicuous white spot on each side of the fifth, sixth, and seventh segments.

The date of the appearance of the caterpillar is in September.

It feeds on the bramble.

NOCTUA BATIS.

PEACH BLOSSOM.

Plate XL. *Figure* 2.

Localities for this species are York, Beverley, Brighton, Sutton-on-Derwent, Charmouth, Faversham, Darenth Wood, Canterbury, Black Park, Dorking, Arundel, the New Forest, Barnstaple, West Looe, Gloucester, Witney, Monks Wood, Rotherham, Preston, Carlisle, Dumfries, Kilmun, Stirling, Killarney.

The situations where it is found are woods.

The perfect insect appears in June and July.

The caterpillar is reddish grey, mottled with brown, with a rather large prominence on the back of the third segment, and a smaller one on the sixth, seventh, eighth, ninth, and tenth segments.

The date of the appearance of the caterpillar is September.

It feeds on the bramble.

CYMATOPHORA DUPLARIS.

LESSER SATIN CARPET.

Plate XL. *Figure* 3.

Localities for this species are York, Beverley, Darenth Wood, Ripley near London, Plymouth, Brighton, Sheffield, Petworth, the New Forest, Epping, Lewes, Preston, West Wickham, Dingwall, Worcester, Worthing, Lyndhurst, Stowmarket, Tenterden, Darlington, Lower Guiting, Halton, Huddersfield, Burton-on-Trent, Scarborough, Birkenhead.

The situations where it is found are woods.

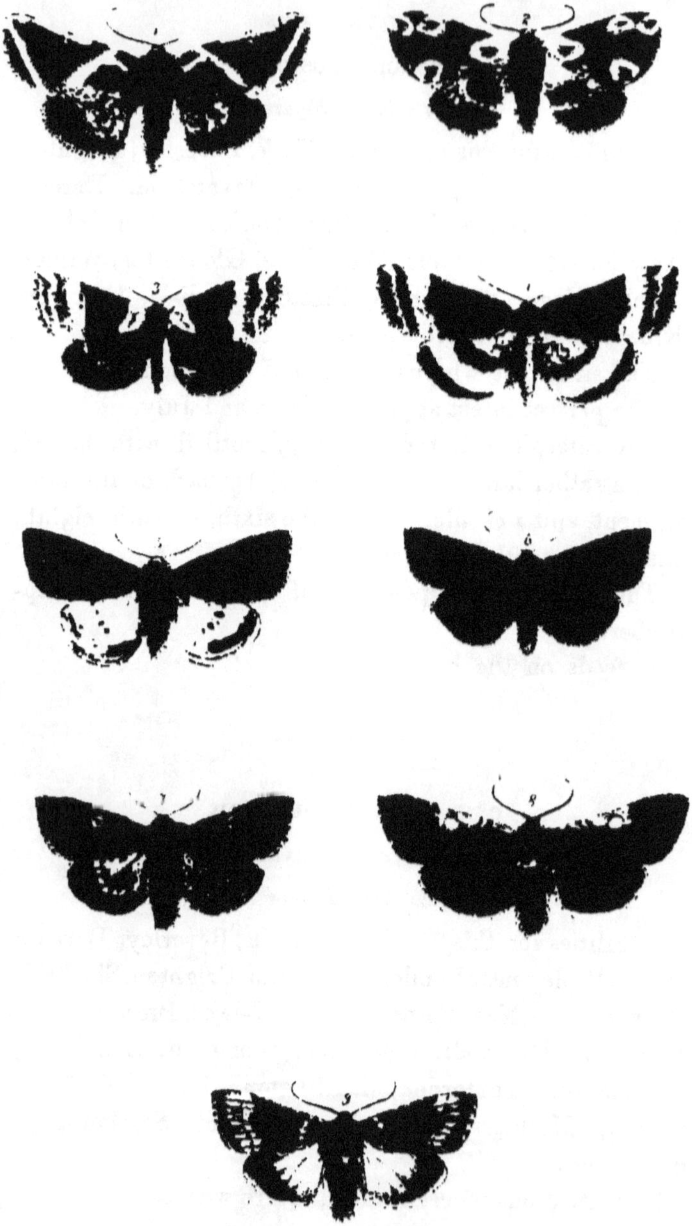

The perfect insect appears in June and July.

The caterpillar is bluish grey, with a row of white dots on each side of the back; below whitish.

The date of the appearance of the caterpillar is in August and September.

It feeds on the birch.

CYMATOPHORA FLUCTUOSA.

SATIN CARPET.

Plate XL. Figure 4.

Localities for this species are Huddersfield, Sheffield, Rotherham, Brighton, Darenth Wood, West Wickham, Pembury, Poyning's, Worcester, Killarney, Tenterden, Stowmarket, Exeter.

The situations where it is found are the outskirts of woods.

The perfect insect appears in June.

The caterpillar is yellowish white, the head blackish brown.

The date of the appearance of the caterpillar is in September and October.

It feeds on the birch.

CYMATOPHORA DILUTA.

LESSER LUTESTRING.

Plate XL. Figure 5.

Localities for this species are York, Doncaster, Sutton-on-Derwent, Brighton, Birch Wood, Arundel, Lyndhurst, Manchester, Bristol, Worcester, Stowmarket, Worthing, Burton-on-Trent, Tenterden, Huddersfield, Wavendon.

The situations where it is found are woods.

The perfect insect appears in August and September.

The caterpillar is grey, with some minute white dots, with a blackish line on each side; the sides and under part whitish, with a row of black dots above the legs.

The date of the appearance of the caterpillar is in June. It feeds on the oak.

CYMATOPHORA OR.

POPLAR LUTESTRING.

Plate XL. *Figure* 6.

Localities for this species are York, Canterbury, Lewes, Darenth Wood, West Wickham, Halton, Stowmarket, Bristol, Tenterden, Worthing, Cambridge, Marlow, Lower Guiting, and near London.

The situations where it is found are woods.

The perfect insect appears in June and July.

The caterpillar is pale yellowish green, with a darker line along the back, and a yellowish side line.

The date of the appearance of the caterpillar is July and August.

It feeds on the poplar.

CYMATOPHORA OCULARIS.

FIGURE-OF-80.

Plate XL. *Figure* 7.

Localities for this species are Netley, Bristol, Brandon, Birmingham, Stowmarket, Worcester, Cambridge, Halton.

The perfect insect appears at the end of June and the beginning of July.

The caterpillar is very pale yellowish green, with a greenish line along the back, and another on the sides.

The date of the appearance of the caterpillar is in August and September.

It feeds on the aspen.

CYMATOPHORA FLAVICORNIS.

YELLOW-HORNED.

Plate XL. *Figure* 8.

Localities for this species are Langwith near York, Brighton, Tilgate Forest, Dulwich, Horndean, Newnham, Black Park, Rugeley, Bristol, Dunham Park, Rannock, Birkenhead, Sheffield, Huddersfield, Lewes, Manchester, Worthing, and near London.

The situations where it is found are in and near woods.

The perfect insect appears in March.

The caterpillar is whitish or dull greenish, with whiter dots, pale brownish between the segments, and with a row of black dots on the sides.

The date of the appearance of the caterpillar is in July, August, and September.

It feeds on the birch.

CYMATOPHORA RIDENS.

FROSTED GREEN.

Plate XL. *Figure 9.*

Localities for this species are York, Brighton, Shooter's Hill, Black Park, Great Torrington, Bristol, Carlisle, Cockermouth, Exeter, Lewes, Lyndhurst, Manchester, Brentwood, the New Forest, Sherwood Forest, Stowmarket, Worcester, Boar's Wood, Kole near Liverpool, and in places near London.

The situations where it is found are woods.

The perfect insect appears in April.

The caterpillar is dark bluish grey, with conspicuous whitish dots, and along the back a plain line of the ground colour, on the sides and the under part whitish. Sometimes it is yellowish, with several green lines, or dull bluish, with several black lines.

It feeds on the oak.

BRYOPHILIDÆ.

BRYOPHILA GLANDIFERA.

MOTTLED GREEN.

Plate XLI. *Figure 1.*

Localities for this species are Brighton, Barnstaple, Falmouth, Bristol, Exeter, Plymouth, Worcester.

The perfect insect appears in July and August.

The caterpillar is green, with a broad dark olive-green stripe along the back, within which is an interrupted white line; the head black.

The date of the appearance of the caterpillar is in February, March, and April.

It feeds on lichens.

BRYOPHILA ALGÆ.

Plate XLI. *Figure* 2.

Localities for this species are near Manchester.
The perfect insect appears in July.

BRYOPHILA PERLA.

MARBLED BEAUTY.   SMALL BISHOP MOTH.

Plate XLI. *Figure* 3.

Localities for this species are York, Sheffield, Sand-Hutton near York, Brighton, Newhaven, Peterborough, Falmouth, Bromsgrove, Birkenhead, Lower Guiting, Huddersfield, Bristol, Darlington, Lewes, Manchester, Exeter, Shrewsbury, Worthing, Burton-on-Trent, Scarborough, Tenterden, Cambridge, Edinburgh, Stowmarket, Worcester.

The perfect insect appears in July and August.

The caterpillar is bluish black, with a broad orange stripe along the back; the head bright black.

The date of the appearance of the caterpillar is in March, April, and May.

It feeds on lichens.

BOMBYCOIDÆ.

DIPHTHERA ORION.

SCARCE MARVEL-DE-JOUR.

Plate XLI. *Figure* 4.

Localities for this species are Brighton, Whippingham in the Isle of Wight, Alverstone, Bere Forest, the New Forest, Hooton near Birkenhead, Darlington, Lyndhurst, Tenterden, Worthing.

The situations where it is found are woods.

The perfect insect appears in May and June. May 24, June 25.

The caterpillar is of a "reddish or yellowish grey, with curved silky hairs; the back black, interrupted by large oval spots of pale yellow."

The date of the appearance of the caterpillar is in August and September.

It feeds on the oak.

The chrysalis is found in a cocoon.

ACRONYCTA TRIDENS.

DARK DAGGER.

Plate XLI. *Figure* 5.

Localities for this species are York, Nunburnholme, Nafferton, Brighton, Bromsgrove, Sutton-on-Derwent, Bristol, Huddersfield, Shrewsbury, Worthing, Cambridge, Lewes, Stowmarket, Exeter, Worcester, Halton.

The situations where it is found are hedge sides, gardens, woods, &c.

The perfect insect appears in June.

The caterpillar is orange: red on the upper part, with a small black prominence on the fifth segment, and a white one tipped with black on the twelfth.

The date of the appearance of the caterpillar is in August and September.

It feeds on various plants and shrubs.

The chrysalis is found in a cocoon amongst moss, or in crevices of bark.

ACRONYCTA PSI.

DAGGER MOTH.   COMMON DAGGER.

Plate XLI. *Figure* 6.

Localities for this common species are York, Nunburnholme, Nafferton, Sutton-on-Derwent, Brighton, Anstey, Bromsgrove.

The situations where it is found are woods, gardens, hedge sides, &c.

The perfect insect appears in June, July, and August.

The caterpillar is greyish black, with a broad pale yellow line along the back, a large black prominence on the fifth segment, and a short black one on the twelfth.

The date of the appearance of the caterpillar is in August, September, and October.

It feeds on various plants.

The chrysalis is found in a cocoon amongst moss, or in crevices of bark.

ACRONYCTA LEPORINA.

MILLER.

Plate XLI. *Figure* 7.

Localities for this species are York, Rotherham, Pembury, Askham Bog, Langwith, Brighton, Canterbury, St. Osyth's, West Wickham, Balcombe, Manchester, Bristol, Plymouth, Bere Forest, Sherwood Forest, Huddersfield, Stowmarket, Crewe, Lewes, Tenterden, Worthing, Derby, Lyndhurst, Preston, Glasgow, Carlisle, and near London.

The situations where it is found are woods.

The perfect insect appears in June and July.

The caterpillar is pale green, thickly covered with long white hairs, and a few dark ones on the second, third, fourth, eleventh, and thirteenth segments.

The date of the appearance of the caterpillar is in August and September.

It feeds on the birch.

The chrysalis is found in a cocoon among moss or in crevices of bark.

ACRONYCTA ACERIS.

SYCAMORE MOTH.

Plate XLI. *Figure* 8.

Localities for this species are Faversham, Brighton, Lewisham, Cambridge, Tenterden, Birchwood, Lewes, Worthing, Dulwich, Stowmarket, Newhaven, Barnstaple, Black Park, Sudbury, Peterborough.

The perfect insect appears in June.

The caterpillar is yellowish, with long tufts of orange hair, and an angular-shaped white spot edged with black on the back of each segment.

The date of the appearance of the caterpillar is in August.

It feeds on the sycamore and the horse-chesnut.

The chrysalis is found in a cocoon among moss or in crevices of bark.

ACRONYCTA MEGACEPHALA.

POPLAR GREY.

Plate XLII. *Figure* 1.

Localities for this species are York, Brighton, the Isle of Man, Lewisham, Taunton, Cambridge, Canterbury, Manchester.

The perfect insect appears in June and July.

The caterpillar is yellowish grey, dotted with black along the back, and with reddish raised spots, the tenth segment with a large pale blot.

The date of the appearance of the caterpillar is in August.

It feeds on the poplar.

The chrysalis is found in a cocoon among moss or in crevices of bark.

ACRONYCTA STRIGOSA.

Plate XLII. Figure 2.

Localities for this species are Monkswood, and Fulbourn near Cambridge.

The perfect insect appears in July; July 16.

The caterpillar is green on the sides, with a reddish brown stripe along the back, dotted with black towards the sides, and edged with yellowish; a slight eminence on the back of the fifth and twelfth segments.

The date of the appearance of the caterpillar is in September.

It feeds on the sloe.

The chrysalis is found in a cocoon among moss or in crevices of bark.

ACRONYCTA ALNI.

ALDER MOTH.

Plate XLII. Figure 3.

Localities for this species are Boxhill, York, Doncaster, Brantingham, Charford near Bromsgrove, Brighton, Wakefield, Holme Bush in Sussex, Brighthampton, Morton near

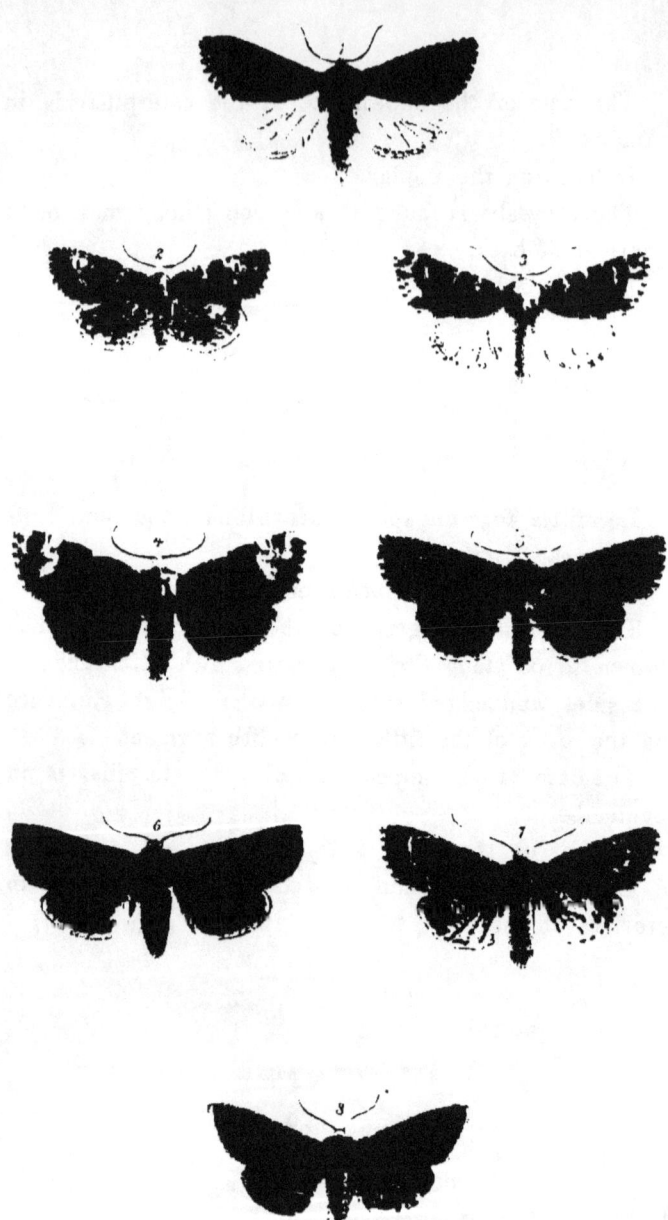

Gainsborough, Halton, Hastings, Burton-on-Trent, Huddersfield, Lewes, Manchester, Ryde, Rugby, Lyndhurst, Pembury, Worthing, Worcester, Sheffield, Shrewsbury, Wavendon, Wentworth near Rotherham, Howsham, Scarborough, Liverpool, Colnethorpe Wood near Witney, and Speke Hall near Liverpool.

The situations where it is found are gardens, woods, &c.

The perfect insect appears in May, June, July, and August; May 27, 28; June 6, 12; August 4, 24. It has also been known to emerge from the chrysalis in September; September 8.

The caterpillar is purple black, with a square spot on the back of the second, third, fourth, fifth, sixth, seventh, eighth, ninth, tenth, eleventh, and twelfth segments, and two long black hairs or spines, with a spade-shaped club or tip on each segment.

The date of the appearance of the caterpillar is in July, August, and September; July 27, 30; August 1, 11.

It feeds on the alder.

The chrysalis is found in a cocoon among moss or in crevices of bark.

ACRONYCTA LIGUSTRI.

CORONET.

Plate XLII. *Figure* 4.

Localities for this species are York, Nafferton, Stockton Forest, Brighton, Bere Forest, Ventnor, Birch Wood, West Looe, Bristol, Cambridge, Exeter, Darenth Wood, West Wickham, Malvern, Peterborough, Black Park, Doncaster,

Lewisham, Lower Guiting, Halton, Canterbury, Carlisle, Lewes, Manchester, Plymouth, Stowmarket, Sevenoaks, Lyndhurst, Pembury, Tenterden, Worthing, Worcester, Newhaven, Swinhope, Speke, Hale, and Rainhill near Liverpool.

The situations where it is found are woods and gardens.

The perfect insect appears in June and July.

The caterpillar is green, with a white line and spots along the back, the side line yellow, with red dots.

The date of the appearance of the caterpillar is in August.

It feeds on the ash and the privet.

The chrysalis is found in a cocoon among moss or in crevices of bark.

ACRONYCTA RUMICIS.

BRAMBLE MOTH.

Plate XLII. *Figure* 5.

Localities for this common species are York, Sutton-on-Derwent, Askham Bog, Brighton, Derby, Bognor, Black Park, Faversham, Carlisle, Manchester, Dunoon, Stirling.

The situations where it is found are woods, etc.

The perfect insect appears in May, June, and July.

The caterpillar is blackish, each segment black in front, with a white spot on the sides and an orange one in the middle of each; the side line white, spotted with red.

The date of the appearance of the caterpillar is in August and September.

It feeds on various plants.

The chrysalis is found in a cocoon among moss or in crevices of bark.

ACRONYCTA AURICOMA.

SCARCE DAGGER.

Plate XLII. *Figure 6.*

Localities for this species are Canterbury, Brighton, Tenterden, and near London.

The situations where it is found are woods.

The perfect insect appears in July and August.

The caterpillar is purple grey, with reddish orange spots on the back, a white stripe along the sides, with black raised dots, and a row of the like above them.

The date of the appearance of the caterpillar is in September.

It feeds on the bramble, the bilberry, the birch, etc.

The chrysalis is found in a cocoon amongst moss or in crevices of bark.

ACRONYCTA MENYANTHIDIS.

LIGHT KNOT-GRASS.

Plate XLII. *Figure 7.*

Localities for this species are York, Saddleworth, Carlisle, Glasgow, Huddersfield, Manchester, Sheffield, Stowmarket.

The situations where it is found are heaths and moors

The perfect insect appears in June and July.

The caterpillar is black, with a broad red stripe on each side above the feet.

The date of the appearance of the caterpillar is in September.

It feeds on the heather and the sweet gale, (*Myrica Gale*).

The chrysalis is found in cocoon in crevices of back or among moss.

ACRONYCTA MYRICÆ.

SPURGE MOTH.

Plate XLII. *Figure* 8.

Localities for this species are the New Forest (?) and Rannock.

The situations where it is found are the moors and heathy places.

The perfect insect appears in May and June; May 27, June 15.

The date of the appearance of the caterpillar is in August and September.

It feeds on the sweet gale *(Myrica Gale)*, and the sallow *(Salix capræa)*.

The chrysalis is found in a cocoon in crevices of bark or in moss.

NOCTUÆ.

SIMYRA VENOSA.
POWDERED WAINSCOT.

Plate XLIII. *Figure* 1.

Localities for this species are Cambridge, Falmouth, and Stowmarket.

The situations where it is found are the Fens and watery places.

The perfect insect appears in June.

The caterpillar is whitish, with a blackish stripe along the back, and a dark grey one on the sides, with raised yellow spots on the back and sides.

The date of the appearance of the caterpillar is in September.

It feeds on the reed meadow-grass (*Poa aquatica*), and other water plants.

The chrysalis is enclosed in a cocoon of silk.

GENUINÆ.—LEUCANIDÆ.

SYNIA MUSCULOSA.

Plate XLIII. *Figure* 2.

Localities for this species are Darenth Wood, Brighton.
The perfect insect appears in August; August 7.

LEUCANIA CONIGERA.

BROWN LINE. BRIGHT EYE.

Plate XLIII. *Figure* 3.

Localities for this species are Sutton-on-Derwent, Nunburnholme, Charmouth, York, Scarborough, Darlington, the Isle of Man, Brighton, Malvern, Dorking, Bristol, Cambridge, Edinburgh, Halton, Lewes, Barnstaple, Lower Guiting, Huddersfield, Manchester, Plymouth, Worcester, Perth, Isle of Man, New Brighton, and Dacre Park near Birkenhead, as also near London.

The situations where it is found are woods.

The perfect insect appears in June and July.

The caterpillar is whitish yellow, with a darker line along the back, and a dark brown line on either side below it, underneath which is successively a whitish, a brownish, and then another whitish line.

The date of the appearance of the caterpillar is in April. (?)

It feeds on grass.

This species sometimes flies by day in the bright sunshine.

LEUCANIA VITELLINA.

Plate XLIII. *Figure* 4.

Localities for this species are Brighton, Isle of Wight. The perfect insect appears in August and September.

The date of the appearance of the caterpillar is in October, November, December, January, February, and up to March.

It feeds on grass.

LEUCANIA TURCA.

THE DOUBLE-LINE.

Plate XLIII. Figure 5.

Localities for this species are Rotherham, Lyndhurst, the New Forest, Coombe Wood, Black Park, Bristol, Lewes, Manchester.

The situations where it is found are woods.

The perfect insect appears in June and July.

The caterpillar is marbled, yellowish and grey, with a whitish line along the back, and a series of obscure angular-shaped marks, paler on each side.

The date of the appearance of the caterpillar is in February and March.

It feeds on grasses, in woods, especially on the wood rush (*Luzula Campestris*).

LEUCANIA LITHARGYRIA.

THE OCHRACEOUS BROWN.

Plate XLIII. Figure 6.

Localities for this common species are York, Nunburnholme, Sutton-on-Derwent, Brockenhurst, and other places in the New Forest, Charmouth, Brighton, Faversham, Cambridge, Preston, Birkenhead, Dulwich, Lewisham, Black Park, Arundel, Barnstaple.

The situations where it is found are gardens, hedgesides, etc.

The perfect insect appears in July.

The caterpillar is whitish, with a darker line along the back, and three rather broad ones on each side, the middle one darker than the others.

The date of the appearance of the caterpillar is in March.

It feeds on the chickweed (*Alsine media*), and the plantain (*Plantago lanceolata*), etc.

LEUCANIA EXTRANEA.

Plate XLIII. *Figure* 7.

Localities for this species are the Isle of Wight, Lewes. The perfect insect appears in September; September 9.

LEUCANIA OBSOLETA.

OBSCURE WAINSCOT.

Plate XLIII. *Figure* 8.

Localities for this species are Hammersmith Marshes, Bermondsey Marshes, Fairbrook, Faversham, Dorking, Peterborough, and Bidston, near Birkenhead.

The situations where it is found are Fens and watery places among reeds.

The perfect insect appears in June.

The caterpillar is yellowish grey, with a tinge of rose-colour; the line along the back whitish, edged with dark green, the side line below it narrow and whitish, and another pale line on the lower part of the side.

The date of the appearance of the caterpillar is in August and September.

It feeds on the reed *(Arundo phragmites.)*

LEUCANIA PUTRESCENS.

Plate XLIII. *Figure* 9.

Localities for this species are Torquay and Teignmouth.
The perfect insect appears in July and August; Aug. 27.

LEUCANIA LITTORALIS.

SHORE WAINSCOT.

Plate XLIV. *Figure* 1.

Localities for this species are Christchurch, Deal, Waterford, Lytham, the Isle of Wight, Birkenhead, Bristol.

The situations where it is found are sand hills on the sea coast.

The perfect insect appears in July.

LEUCANIA PUDORINA.

STRIPED WAINSCOT.

Plate XLIV. *Figure* 2.

Localities for this species are York, Askham Bog, Scarborough, the New Forest, Whittlesea Mere, Brighton, Cambridge, Bidston near Birkenhead.

The situations where it is found are the Fens and marshy and boggy places.

The perfect insect appears in July.

The caterpillar is yellowish grey, with a whitish line along the back, and one black line and two rows of black dots on each side.

The date of the appearance of the caterpillar is in March and April.

It feeds on several kinds of grass.

LEUCANIA COMMA.

SHOULDER-STRIPED WAINSCOT.

Plate XLIV. *Figure* 3.

Localities for this species are York, Askham Bog, Sutton-on-Derwent, Huddersfield, Nunburnholme, Black Park, Brighton, Lewisham, Bere Forest, Barnstaple, Chester, Birkenhead, Bristol, Cambridge, Lower Guiting, Burton-on-Trent, Manchester, Scarborough, Shrewsbury, Lewes, Worthing, Tenterden, Worcester.

The situations where it is found are woods, gardens, moist places, brook-sides, etc.

Plate XLIV

The perfect insect appears in June and July.

The caterpillar is reddish brown, with three rows of black dots on each side.

The date of the appearance of the caterpillar is in September and October, and on to the spring.

It feeds on the sorrel *(Rumex acetosella)* and various grasses.

LEUCANIA STRAMINEA.

SOUTHERN WAINSCOT.

Plate XLIV. *Figure 4.*

Localities for this species are the Isle of Wight, Hammersmith Marshes, Whittlesea Mere, Poynings, Leasowe and Meols near Birkenhead.

The situations where it is found are fens and marshy places.

The perfect insect appears in June and July.

The caterpillar is dull yellowish red, with a line along the back, and two rows of blackish dots on each side of it, with numerous short lines on the sides, black and pale dull yellowish red alternately.

The date of the appearance of the caterpillar is in February, March, and April.

It feeds on several grasses.

LEUCANIA IMPURA.

SMOKY WAINSCOT.

Plate XLIV. *Figure 5.*

Localities for this species are York, Askham Bog, Nunburnholme, Brighton, Faversham, Dorking, Isle of Man, Poynings, Barnstaple, Uppingham.

The situations where it is found are gardens, hedge-sides, etc.

The perfect insect appears in June, July, and August.

The caterpillar is yellowish grey, with a white line along the back, and a whitish one below it on either side, its upper edge black.

The date of the appearance of the caterpillar is in March, April, and May.

It feeds on the sedge *(Carex divisa)*, etc.

LEUCANIA PALLENS.

COMMON WAINSCOT.

Plate XLIV. *Figure* 6.

Localities for this species are York, Askham Bog, Sutton-on-Derwent, Nunburnholme, Brighton, Dorking, Faversham, Barnstaple, the Isle of Man.

The situations where it is found are gardens, hedge-sides, etc.

The perfect insect appears in June, July, and August.

The caterpillar is greyish with a tinge of dull yellowish red, a line along the back with another below it on each side edged above with black, underneath which are three stripes, one reddish, one grey, and one dull yellowish red.

The date of the appearance of the caterpillar is in March and April.

It feeds on different grasses.

LEUCANIA PHRAGMITIDIS.

Plate XLIV. *Figure* 7.

Localities for this species are Deal, Greenwich Marshes, and the Cambridgeshire Fens.

The situations where it is found are the fens and reedy places.

The perfect insect appears in July.

The caterpillar is dull white with a row of large irregular blue brown spots on each side, the head and the hind segments shining black.

The date of the appearance of the caterpillar is in May.

It feeds on the young stems of the Reed (*Arundo phragmites*).

MELIANA FLAMMEA.

Plate XLIV. *Figure* 8.

Localities for this species are the Cambridgeshire Fens.
The situations where it is found are the Fens.
The perfect insect appears in June.

SENTA ULVÆ.

Plate XLIV. *Figure* 9.

Localities for this species are Hackney Marshes, Hammersmith Marshes, and the Cambridgeshire Fens.

The situations where it is found are fens and marshy places.

The perfect insect appears in August.

The caterpillar is dull yellowish, with several fine lines.

The date of the appearance of the caterpillar is in October, March, and April.

It feeds in the reed (*Arundo phragmites*).

The chrysalis is found enclosed in the stems of reeds.

NONAGRIA DESPECTA.

LINEATED RUFOUS.

Plate XLIV. *Figure* 10.

Localities for this species are near Cambridge.

The situations where it is found are fens and marshy places.

The perfect insect appears in June and July.

NONAGRIA FULVA.

Plate XLV. *Figure* 1.

Localities for this species are York, Askham Bog, Sutton-on-Derwent, Whittlesea Mere, Clapham Common, Bristol, Brighton, Birkenhead, Edinburgh, Kingsbury, Darlington, Scarborough, Burton-on-Trent, Manchester.

The situations where it is found are woods, marshy places, and fens.

The perfect insect appears in August and September; September 15.

The caterpillar is dull white, with a reddish stripe along the back, and a blackish line on the sides, over a row of black raised dots.

Plate XLIV.

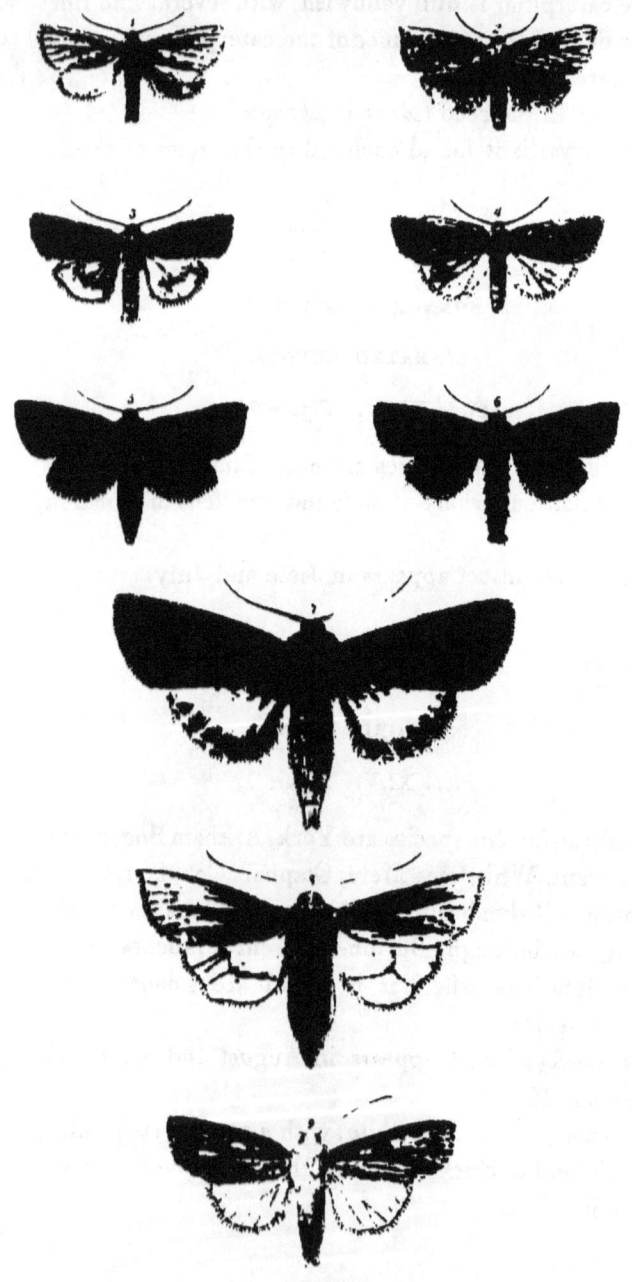

The date of the appearance of the caterpillar is in May and June.

It feeds in the stems of the reed meadow-grass *(Poa aquatica)*, and the sedge *(Carex divisa)*.

The chrysalis is found in the stem of the plant on which the caterpillar has fed.

NONAGRIA CONCOLOR.

Plate XLV. Figure 2.

Localities for this species are Folkestone, Worcester, Whittlesea Mere, and the Fens of Cambridgeshire.

The situations where it is found are marshy places and fens.

The perfect insect appears in June.

The chrysalis is found within the stem of the plant on which the caterpillar has fed.

NONAGRIA HELLMANNI.

Plate XLV. Figure 3.

Localities for this species are Yaxley Fen and Whittlesea Mere.

The situations where it is found are fenny districts.

The perfect insect appears in June.

NONAGRIA NEURICA.

Plate XLV. *Figure* 4.

Localities for this species are Whittlesea Mere and Yaxley Fen.

The situations where it is found are fens and fenny places.

The perfect insect appears in July and August.

The caterpillar is dull white, with a pale red line along the back.

The date of the appearance of the caterpillar is in April and May.

It feeds on the stem of the reed (*Arundo phragmites*).

The chrysalis is found in the stem of the reed.

NONAGRIA GEMINIPUNCTA.

TWIN-SPOT WAINSCOT.

Plate XLV. *Figure* 5.

Localities for this species are Weston-super-Mare, the Hammersmith Marshes, and the Cambridgeshire Fens.

The situations where it is found are fens and marshes.

The perfect insect appears in August.

The caterpillar is dull white dotted with brownish, the head shining brownish red, the raised spots on the side black.

The date of the appearance of the caterpillar is in May.

It feeds on the stem of the reed (*Arundo phragmites*).

The chrysalis is found in the stem of the reed.

NONAGRIA CANNÆ.

Plate XLV. *Figure* 6.

Localities for this species are in Yaxley Fen.

The situations where it is found are the fens.

The perfect insect appears in August.

The caterpillar is greenish or yellowish, with black dots, the head brownish, the spots on the side black.

The date of the appearance of the caterpillar is in May.

It feeds on the reed mace (*Typha latifolia*).

The chrysalis is found in the stem of the food-plant of the caterpillar.

NONAGRIA TYPHÆ.

BULLRUSH MOTH.

Plate XLV. *Figure* 7.

Localities for this species are York, Buttercrambe Moor, Birkenhead, Scarborough, Brighton, Hammersmith, Stowmarket, Burton-on-Trent, Cambridge, Plymouth, Huddersfield, Manchester, Worthing, Kingsbury, Wavendon.

The situations where it is found are marshy places.

The perfect insect appears in August, September, and October. September 10.

The caterpillar is dull reddish yellow, with a pale line along the back, the head yellowish brown, the raised spot on the sides blackish.

The date of the appearance of the caterpillar is in May and June.

It feeds on the reed mace (*Typha latifolia*).

The chrysalis is found within the stem of the plant.

NONAGRIA LUTOSA.

Plate XLV. *Figure* 8.

Localities for this species are York, Scarborough, Huddersfield, Croydon, Hammersmith, Brighton, Worthing, Ely, Plumstead, Marlow, Preston, Birkenhead, Bristol, Burton-on-Trent.

The situations where it is found are fens and marshy places, where reeds grow.

The perfect insect appears at the end of August, and in September, October, and to the beginning of November; September 10, October 19, October 21.

The date of the appearance of the caterpillar is in June and July.

It feeds in the roots of the reed *(Arundo phragmites)*.

NONAGRIA ELYMI.

Plate XLV. *Figure* 9.

Localities for this species are the Norwich Fens.

The situations where it is found are fens and such places.

The perfect insect appears in June; June 27.

APAMIDÆ.

GORTYNA FLAVAGO.

FROSTED ORANGE. BURDOCK MOTH.

Plate XLVI. *Figure* 1.

Localities for this species are York, Scarborough, Brigh-

Plate XLVI

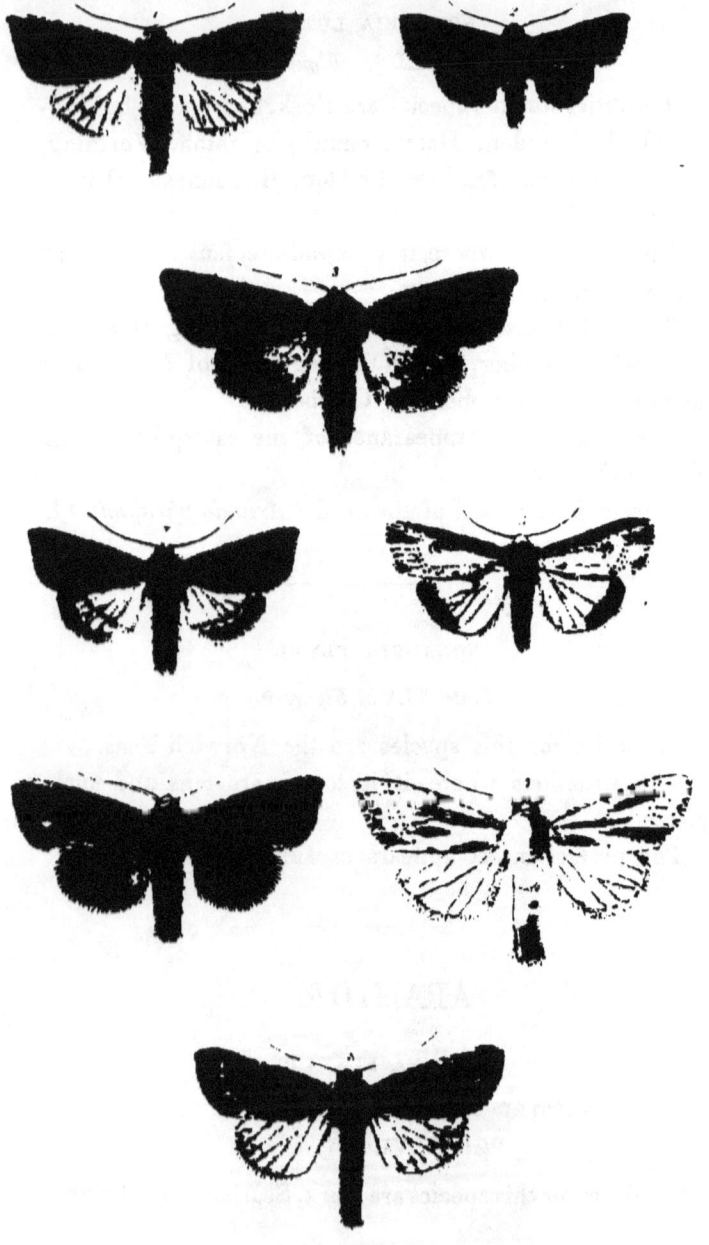

ton, Sherwood Forest, Exeter, Huddersfield, Shrewsbury, Bromsgrove, Birkenhead, Bristol, Burton-on-Trent, Cambridge, Lower Guiting, Darlington, Kingsbury, Plymouth, Manchester, Worcester.

The situations where it is found are lanes, hedge-sides, etc.

The perfect insect appears in September.

The caterpillar is pale dull yellow with conspicuous black dots, the hind segment tinged with greyish brown.

The date of the appearance of the caterpillar is in June.

It feeds in the stem of the burdock (*Arctium lappa*), the water betony (*Scrophularia aquatica*), the thistle, etc.

The chrysalis is found within the stem of the plant.

This species is attracted by light.

HYDRÆCIA NICTITANS.

GOLDEN BAR.

Plate XLVI. *Figure* 2.

Localities for this species are York, Sutton-on-Derwent, Nunburnholme, Margate, Birch Wood, Putney Heath, the New Forest, Brighton, the Isle of Man, Lullingstone Park, Lewisham, Black Park, Saddleworth, Preston, Birkenhead, Lyndhurst, Exeter, Worthing, Scarborough, Tenterden, Plymouth, Manchester, Kingsbury, Edinburgh, Dunoon, Cambridge, Darlington, Bristol, Huddersfield, Halton, and the Isle of Arran.

The situations where it is found are gardens, woods, etc.

The perfect insect appears in July, August, and September; August 4, August 7, September 3.

The caterpillar is dull brown, with a line along the back, and a row of brown dots on each side of it.

The date of the appearance of the caterpillar is in May and June.

It feeds on the roots of different grasses.

The chrysalis is found in a cocoon of earth.

This species also comes to a light.

HYDRÆCIA PETASITIS.

Plate XLVI. *Figure* 3.

Localities for this species are Carron, Stirling, Carlisle, Failsworth, Edinburgh, Manchester, Liverpool.

The situations where it is found are the banks of streams.

The perfect insect appears in August.

The caterpillar is dull whitish, with black dots.

The date of the appearance of the caterpillar is in June and July.

It feeds on, or rather in, the stems and roots of the burdock (*Arctium lappa*), and the butter bur (*Petasites vulgaris*).

The chrysalis is found in a cocoon of earth.

HYDRÆCIA MICACEA.

ROSY RUSTIC.

Plate XLVI. *Figure* 4.

Localities for this species are York, Sutton-on-Derwent, Nunburnholme, Bromsgrove, the Isle of Man, Brighton, Lewisham, Carron, Stirling, Scarborough.

The situations where it is found are near moist and watery ground and ditches, also in gardens and lanes, and by hedge-sides.

The perfect insect appears in August and September; August 28.

The caterpillar is pale dull yellowish red, with black dots.

It feeds on the roots of various plants.

The chrysalis is found in an earthen cocoon.

AXYLIA PUTRIS.

FLAME. HORSE-RADISH MOTH.

Plate XLVI. *Figure* 5.

Localities for this species are York, Sutton-on-Derwent, Humberstone, Brighton, Faversham, Barnstaple, Lewes, Tenterden, Worcester, Black Park, Newhaven, Sudbury, Falmouth, Stowmarket, Scarborough, Plymouth, Carlisle, Shrewsbury, Manchester, Kingsbury, Birkenhead, Southport, Bristol, Burton-on-Trent, Cambridge, Edinburgh, Lower Guiting, Darlington, Exeter, Halton.

The situations where it is found are gardens, woods, &c.

The perfect insect appears at the end of June and in July.

The caterpillar is brown, with a yellowish line along the back; with one yellow and two white dots on each segment, and a triangular-shaped black blot on the fifth and sixth.

The date of the appearance of the caterpillar is in August.

It feeds on various low plants.

The chrysalis is found beneath the surface of the earth.

XYLOPHASIA RUREA.

CLOUDED BORDERED BRINDLE.

Plate XLVI. *Figure* 6.

Localities for this very common species are York, Nunburnholme, Sutton-on-Derwent, Charmouth, Brighton, Darenth Wood, Faversham, the New Forest, Black Park, Preston, Edinburgh, Falmouth.

The situations where it is found are gardens, hedge-sides, etc.

The perfect insect appears in June and July.

The caterpillar is dark brownish shining red, with a white line along the back edged with brown, and a brown streak on the sides, bordered on its upper part with red.

The date of the appearance of the caterpillar is in March and April.

It feeds on the primrose (*Primula vulgaris*), the sorrel (*Rumex acetosella*), and various grasses, etc.

The chrysalis is found enclosed in a slight cocoon of earth.

XYLOPHASIA LITHOXYLEA.

LIGHT ARCHES.

Plate XLVI. *Figure* 7.

Localities for this common species are York, Nunburnholme, Sutton-on-Derwent, Charmouth, Brighton, Faversham, Black Park, Dorking, Peterborough, the Isle of Man, Penzance, and near Bream Bay, Falmouth.

The situations where it is found are gardens, woods, and hedgerows.

The perfect insect appears in June and July.

XYLOPHASIA SUBLUSTRIS.

REDDISH LIGHT-ARCHES.

Plate XLVI. *Figure* 8.

Localities for this species are York, Brighton, Fairbrook, Arundel, Black Park, Monks' Wood, Preston, Bristol, Lewes, Cambridge, Lower Guiting, Plymouth, Scarborough, Stowmarket, Marlow, and near Galway.

The situations where it is found are woods.

The perfect insect appears in June and July.

XYLOPHASIA POLYODON.

DARK ARCHES.

Plate XLVII. *Figure* 1.

Localities for this very abundant species are York, Nunburnholme, Charmouth, Worcester, Isle of Man, Tain, Anstey, Brighton, Falmouth, Birkenhead, Sutton-on-Derwent, etc., etc., etc., etc.

The situations where it is found are gardens, hedgesides, lanes, woods, etc.

The perfect insect appears in June and July.

The caterpillar is dull greyish brown, with bright black spots, the second and last segments black.

The date of the appearance of the caterpillar is in April and May.

It feeds on the roots of grasses and other low plants.

The chrysalis is found beneath the surface, in a slight cocoon of earth.

This species comes much to a light.

XYLOPHASIA HEPATICA.

CLOUDED BRINDLE.

Plate XLVII. *Figure* 2.

Localities for this species are York, Sutton-on-Derwent, Brighton, Darenth Wood, Black Park, Bere Forest, Flint, Kingston in Kent, Lyndhurst, Stowmarket, Worcester, Tenterden, Worthing, Wavendon, Plymouth, Manchester, Bristol, Burton-on-Trent, Darlington, Cambridge, Exeter, Halton, Huddersfield, Kingsbury, Lewes, Hargrave Hall near Birkenhead, Hale and Croxteth near Liverpool.

The situations where it is found are gardens, woods, etc.

The perfect insect appears in June and July.

The caterpillar is dull grey, marbled with darker, and with a whitish line along the back.

The date of the appearance of the caterpillar is in April and May.

It feeds on the roots of several low plants.

The chrysalis is found in a slight cocoon of earth under the ground.

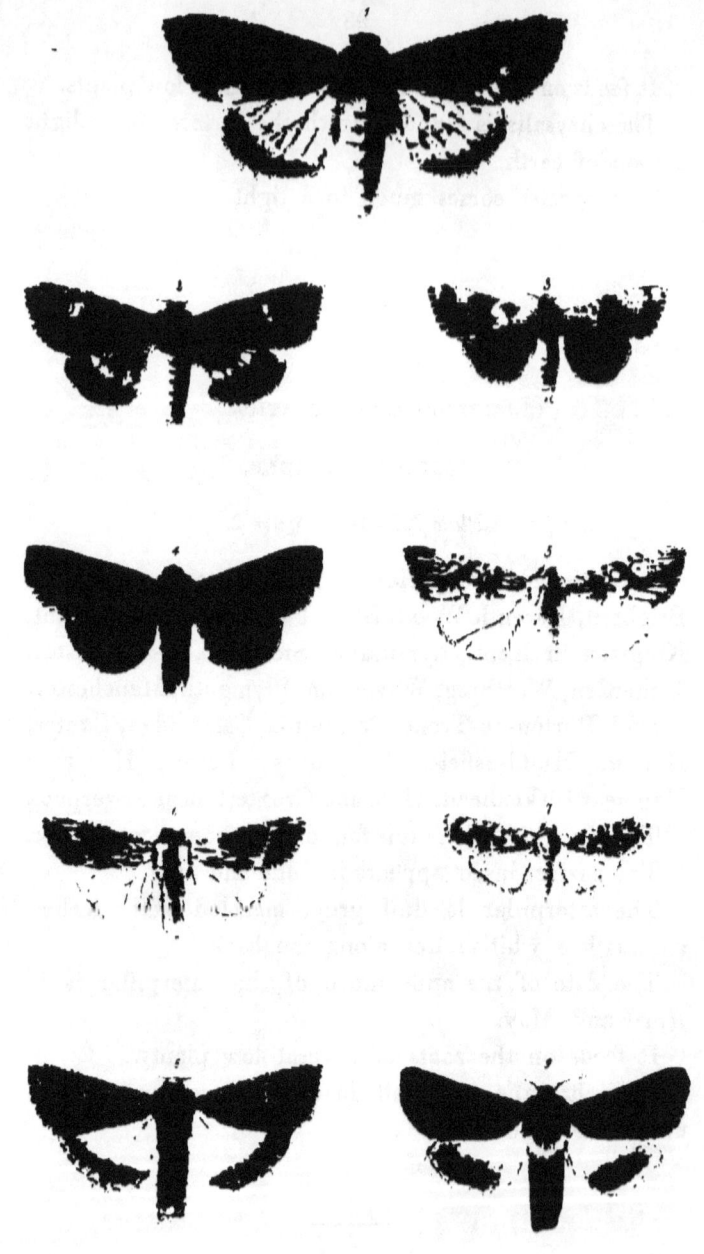

XYLOPHASIA SCOLOPACINA.

SLENDER CLOUDED BRINDLE.

Plate XLVII. *Figure* 3.

Localities for this species are Huddersfield, Sutton-on-Derwent, Carlisle, Lewisham, Barnstaple, Doncaster, Burton-on-Trent, Manchester, Stowmarket, Sherwood Forest, Marlow, Chesterfield, Carlisle, and near London.

The perfect insect appears in July and August.

The caterpillar is bluish grey, with a paler line along the back, and another on each side of it, the lower part of the sides pale yellow.

The date of the appearance of the caterpillar is in May.

It feeds on the rush of different kinds, as the club rush (*Scirpus lacustris*), the quaking grass (*Briza media*), etc.

The chrysalis is found underneath the surface, enclosed in a slight cocoon of earth.

DIPTERYGIA PINASTRI.

BIRDS-WING.

Plate XLVII. *Figure* 4.

Localities for this species are Ampfield, Lewes, Black Park, Lewisham, Worcester, Stowmarket, Kingsbury, Dorking, Manchester.

The situations where it is found are gardens, etc., near pine plantations.

The perfect insect appears in June.

The caterpillar is rich brown, with a darker line along the back, and a broad dull white one on the sides.

It feeds on the sorrel (*Rumex acetosella*).

The chrysalis is found in a cocoon on the surface of the earth.

XYLOMIGES CONSPICILLARIS.

SILVER CLOUD.

Plate XLVII. *Figure* 5.

Localities for this species are Bulstrode Park, Birch Wood, Taunton, Worcester, Darenth Wood.

The situations where it is found are woods.

The perfect insect appears at the end of March, and in April, May, and June; March 27, April 10, April 14, May 3, June 2.

The caterpillar is reddish brown, marbled with brown and whitish, with a broad line of a paler shade on the side, edged above with darker.

The date of the appearance of the caterpillar is in July.

It feeds on the bird's-foot trefoil (*Lotus corniculatus*), and other plants.

The chrysalis is found under the ground.

APOROPHILA AUSTRALIS.

Plate XLVII. *Figure* 6.

Localities for this species are Yarmouth, in the Isle of Wight, Torquay, Lowestoft, Brighton, Deal, Teignmouth, Lewes, Plymouth.

The situations where it is found are sand-hills near the sea.

The perfect insect appears at the beginning of August and in September.

The caterpillar is reddish yellow, with a paler line on the back, the side line powdered with minute spots, and edged above with black spots.

The date of the appearance of the caterpillar is in March.

It feeds on various low plants.

The chrysalis is found underneath the surface.

This species is attracted by light.

LAPHYGMA EXIGUA.

Plate XLVII. *Figure* 7.

Localities for this species are Brighton, Ventnor and Sandown in the Isle of Wight, Worthing.

The perfect insect appears in June.

The chrysalis is subterranean.

NEURICA SAPONARIÆ.

BORDERED GOTHIC.

Plate XLVII. *Figure* 8.

Localities for this species are Stockton Forest near York, Halton, Lewes, Nunburnholme, Swinhope, Black Park, Brighton, Ditton and Rainhill near Liverpool, Worthing, Bere Forest, Burton-on-Trent, Cambridge, Lower Guiting.

The situations where it is found are gardens, woods, and hedges.

The perfect insect appears in June.

The caterpillar is reddish grey, streaked with brown, the side line whitish, the head brown.

The date of the appearance of the caterpillar is in July and August.

It feeds on several low plants, particularly the bladder campion (*Silene inflata*).

The chrysalis is found beneath the surface of the earth.

HELIOPHOBUS POPULARIS.

FEATHERED GOTHIC.

Plate XLVII. *Figure* 9.

Localities for this species are York, Sutton-on-Derwent, Scarborough, Worthing, Nunburnholme, Crambe, Lewes, Brighton, Peterborough, New Brighton, Exeter, Preston, Southport, Shrewsbury, Falmouth, Tenterden, Wavendon, Birkenhead, Burton-on-Trent, Bristol, Cambridge, Lower Guiting, Darlington, Halton, Lyndhurst, Manchester.

The situations where it is found are gardens, fields, etc.

The perfect insect appears in August.

The caterpillar is bronzed brown, darker on the upper part, with dingy white lines.

The date of the appearance of the caterpillar is in April and May.

It feeds at the roots of grass.

The chrysalis is subterranean.

This moth is much addicted to coming to light.

HELIOPHOBUS HISPIDUS.

Plate XLVIII. *Figure* 1.

Localities for this species are Ventnor, Kibworth, Exmouth, Plymouth, the Isle of Portland.

The situations where it is found are sand-hills and rocks on the coast.

The perfect insect appears in September.

The caterpillar is grey, dotted with black, with a line along the back and one beneath it on each side more distinctly dotted.

The date of the appearance of the caterpillar is in November.

It feeds on the plantain, the lettuce, etc.

The chrysalis is found beneath the surface.

CHARÆAS GRAMINIS.

ANTLER MOTH.

Plate XLVIII. *Figure* 2.

Localities for this species are York, Sutton-on-Derwent, Armthorpe, Bidston, Arundel, Storeton, Birkenhead, the Wrekin, Sherwood Forest, Skiddaw, Snowden, West Rasen, New Brighton, Saddleworth, Dunoon, Inverness, Fort Augustus, Brighton, Bristol, Burton-on-Trent, Darlington, Edinburgh, Huddersfield, Lewes, Manchester, Plymouth, Shrewsbury, Scarborough, Stowmarket.

The situations where it is found are lanes, waste places, pastures, etc.

The perfect insect appears at the end of July, in August, and the beginning of September.

The caterpillar is brown or blackish, with yellowish line along the back, and another on the sides of the same colour.

It feeds on the roots of various grasses.

The chrysalis is subterranean.

This species flies much in the suushine.

PACHETRA LEUCOPHÆA.

FEATHERED EAR.

Plate XLVIII. *Figure* 3.

Localities for this species are Bristol, Brighton, Mickleham.

The situations where it is found are woods.

The perfect insect appears in June, towards the end of the month.

The caterpillar is greyish yellow, with a buff yellow line along the back; the head pale shining brown.

The date of the appearance of the caterpillar is in October and through the winter to April.

It feeds in tufts of grass.

The chrysalis is found in a cocoon amongst moss.

CERIGO CYTHEREA.

STRAW UNDERWING. STRAW-COLOURED UNDERWING.

Plate XLVIII. *Figure* 4.

Localities for this neat species are York, Nunburnholme, Burnby, Stockton Forest, Buttercrambe Moor, Bromsgrove,

Brighton, Lewisham, Black Park, Arundel, Darlington, Henfield, Monks-Wood, Peterborough, Sherwood Forest, Carlisle, Birkenhead, Cambridge, Lower Guiting, Halton, Kingsbury, Bristol, Lewes, Lyndhurst, Burton-on-Trent, Manchester, Stowmarket, Wavendon, Worthing, Worcester.

The situations where it is found are clover fields, gardens, wood-sides, &c.

The perfect insect appears in July and August.

The caterpillar is greyish yellow, the sides brownish, of which colour are the three first segments; the lines black.

The date of the appearance of the caterpillar is in September, and so on through the winter to April.

It feeds on grasses.

The chrysalis is found underneath the surface of the ground.

LUPERINA TESTACEA.

LESSER FLOUNCED-RUSTIC.

Plate XLVIII. *Figure* 5.

Localities for this species are York, Brighton, Bristol, Barnstaple, Peterborough, Isle of Man, Falmouth, Birkenhead, Scarborough, Shrewsbury, Halton, Manchester, Burton-on-Trent, Wavendon, Huddersfield, Darlington, Plymouth, Worcester, Exeter, Lewes.

The perfect insect appears in August and September.

The caterpillar is dull yellowish red; the head pale yellowish brown.

The date of the appearance of the caterpillar is in March.

It feeds on the lower part of the stems of grass.
The chrysalis is found enclosed in an earthen cocoon.
This species is another of those which are attracted by light.

LUPERINA DUMERILII.

Plate XLVIII. *Figure* 6.

Localities for this species are Deal, Freshwater in the Isle of Wight, and the Isle of Arran.

The perfect insect appears in August and September.

LUPERINA CESPITIS.

Plate XLVIII. *Figure* 7.

Localities for this species are York, Brighton, Deal, Lewisham, the New Forest, Isle of Wight, Plymouth, New Brighton, Isle of Arran, Birkenhead, Manchester, Lewes, Worthing.

The situations where it is found are lanes.

The perfect insect appears in August and September.

The caterpillar is brown with pinkish white lines; the second and last segments blackish; the head yellowish brown.

The date of the appearance of the caterpillar is in June.

It feeds on different kinds of grass.

The chrysalis is found in a cocoon of earth.

CRYMODES EXULIS.

Plate XLVIII. *Figure* 8.

Localities for this species are near Plymouth, and in the Isle of Arran.

The perfect insect appears in June and July. June 15, July 25.

The caterpillar is of a dingy white colour.

The date of the appearance of the caterpillar is in May, June, July, and August. June 15, August 15.

It feeds on the stems or roots of grasses.

The moth flies in the day-time.

It is attracted by light.

MAMESTRA ABJECTA.

DUSKY NUTMEG.

Plate XLVIII. *Figure* 9.

Localities for this species, which is a rare one, are Nunburnholme, New Brighton, Jackson's Wood near Birkenhead, York, Lewes, the New Forest, Darenth Wood, Holywell, (in Flintshire,) Waterford, St. Osyth's, Cambridge, Gillingham, (in Kent,) Gravesend.

The situations where it is found are gardens, woods, etc.

The perfect insect appears in July.

This moth comes to a light.

MAMESTRA ANCEPS.

DOUBTFUL NUTMEG.

Plate XLVIII. *Figure* 10.

Localities for this species are York, Colchester, Faversham, Brighton, Chester, Barnstaple, Darenth Wood, Black Park, Falmouth, Manchester, Huddersfield, Burton-on-Trent, Lower Guiting, Halton, Bristol, Stowmarket, Cambridge, Edinburgh, Birkenhead, Wavendon, Lewes, Worthing, Worcester.

The situations where it is found are woods.

The perfect insect appears in June.

The caterpillar is pale brown, with three slightly darker streaks; the spots black; the second and last segments black.

The date of the appearance of the caterpillar is in December, January, and February.

The chrysalis is subterranean in an earthen cocoon.

MAMESTRA ALBICOLON.

WHITE COLON.

Plate XLIX. *Figure* 1.

Localities for this species are Deal, Stowmarket, Birkenhead.

The situations where it is found are sand-hills by the sea.

The perfect insect appears at the end of May, and beginning of June; also in August.

The caterpillar is greyish brown marked with darker; the side line pale yellowish; spots white in a circle of black.

The date of the appearance of the caterpillar is in July and August.

The chrysalis is found in an earthen cocoon under the ground.

MAMESTRA FURVA.

DUSKY BROCADE.

Plate XLIX. *Figure* 2.

Localities for this species are Epping, Black Park, Stockport, Exeter, Scarborough, Preston, Llanferras, Kingstown near Dublin, Stowmarket, Arthur's Seat near Edinburgh, and near London.

The situations where it is found are woods, etc.

The perfect insect appears in June, July, and August; June 10.

The caterpillar is clear pale violet brown; the spots bright black; the head bright black.

The date of the appearance of the caterpillar is in March, April, May, and June.

It feeds on the grey hair-grass (*Aira canescens.*).

The chrysalis is found underneath the ground in an earthen cocoon.

MAMESTRA BRASSICÆ.

CABBAGE MOTH.

Plate XLIX. *Figure* 3.

Localities for this exceedingly common species are York, Nunburnholme, Stockton Forest, Buttercrambe Moor,

Charmouth, Nafferton, Brighton, Falmouth, Worcester, and in London, etc., etc., etc.

The situations where it is found are gardens, hedge-sides, woods, etc.

The perfect insect appears at the end of May, and in June and July.

The caterpillar is dark grey, greyish, on green, with a darker line along the back, and an indistinct one on either side below it; the side-line paler.

The date of the appearance of the caterpillar is in October.

It feeds on the cabbage *(Brassica oleracea)*, and other plants.

The chrysalis is found underneath the surface in a slight cocoon of earth.

This species comes much to a light.

MAMESTRA PERSCICARIÆ.

DOT.

Plate XLIX. *Figure* 4.

Localities for this species are Brighton, Walton near Liverpool, Black Park, Canterbury, Cambridge, Chester, Kingsbury, Lyndhurst, Newnham, Lewes, Manchester, Dorking, Brighton, Plymouth, Worthing, Worcester, Burton-on-Trent, Wavendon, Tenterden, Stowmarket.

The perfect insect appears at the end of June and in July.

The caterpillar is pale green, or reddish grey, with a whitish line along the back, and darker marks on the fifth, sixth, seventh, eighth, ninth, tenth, eleventh, and twelfth segments.

The date of the appearance of the caterpillar is in October.

It feeds on the clematis (*Clematis Vitalba*), and various other low plants.

The chrysalis is found under the ground.

APAMEA BASILINEA.

RUSTIC SHOULDER-KNOT.

Plate XLIX. *Figure* 5.

Localities for this common species are York, Brighton, Stockton Forest, Buttercrambe Moor, Falmouth, etc., etc.

The situations where it is found are gardens, hedge-sides, lanes, etc.

The perfect insect appears in June.

The caterpillar is brownish, with a dull yellowish line along the back, and another below it on either side, with a row of black dots between them; the side line whitish, with black raised spots.

The date of the appearance of the caterpillar is in February and March.

It feeds on various low plants.

The chrysalis is found beneath the surface of the ground.

This moth comes to a light.

APAMEA CONNEXA.

UNION RUSTIC.

Plate XLIX. *Figure* 6.

Localities for this species are York, Buttercrambe Moor,

Stockton Forest, Linwood near Barnsley, Sutton-on-Derwent, Sheffield, Chesterfield, Huddersfield, Scarborough.

The situations where it is found are woods.

The perfect insect appears in June and July.

APAMEA GEMINA.

Plate XLIX. *Figure* 7.

Localities for this common species are York, Buttercrambe Moor, Stockton Forest, Wisbeach, Combe Wood, Darenth Wood, Epping, Brighton, St. Osyth's, Barnstaple, Kilburn, Preston, Black Park.

The situations where it is found are woods.

The perfect insect appears in June and July.

The caterpillar is dark grey, with a whitish line along the back, another of the same colour below it on each side, the side-line dull yellowish, with a row of black raised spots, and another of the like colour above them.

The date of the appearance of the caterpillar is in September and October, April and May.

It feeds on grass.

The chrysalis is subterranean.

APAMEA UNANIMIS.

UNIFORM RUSTIC.   SMALL CLOUDED-BRINDLE.

Plate XLIX. *Figure* 8.

Localities for this species are York, Sutton-on-Derwent, Brighton, Birkenhead, Darlington, Bristol, Edinburgh,

Lewes, Stowmarket, Cambridge, Halton, Manchester, Tenterden, Lower Guiting, Kingsbury.

The perfect insect appears in July and August.

The caterpillar is dull grey, with a white line along the back, the side-line whitish, and a row of black dots between these two.

The date of the appearance of the caterpillar is in September, October, March, and April.

It feeds on grass.

The chrysalis is found beneath the surface of the earth.

APAMEA OPHIOGRAMMA.

DOUBLE-LOBED.

Plate XLIX. *Figure* 9.

Localities for this species are Deptford, and Hammersmith Marshes near London.

The situations where it is found are marshy districts where willows grow.

The perfect insect appears in June.

APAMEA FIBROSA.

THE CRESCENT.

Plate XLIX. *Figure* 10.

Localities for this species are York, Askham Bog, Scarborough, Whittlesea Mere, Brighton, Hammersmith, Killarney, Cambridge, Edinburgh.

The situations where it is found are fenny places.

The perfect insect appears in July and August.

The caterpillar is whitish, dull reddish brown along the back, the second segment black.

The date of the appearance of the caterpillar is in April and May.

It feeds in the flower stems of the yellow flag *(Iris pseudacorus)*.

The chrysalis is found below the ground.

APAMEA OCULEA.

COMMON RUSTIC.

Plate L. *Figure* 1.

Localities for this very abundant species are York, Sutton-on-Derwent, London, Brighton, Humberstone, Nunburnholme, Falmouth, Worcester, Bromsgrove, etc., etc.

The situations where it is found are gardens, hedge-sides, woods, etc.

The perfect insect appears in July and August.

The caterpillar is dull grey or greenish white, with a line on each side of the back, and the side-line dull reddish.

The date of the appearance of the caterpillar is in April and May.

It feeds on grasses of different sorts.

The chrysalis is subterranean.

PLATE I.

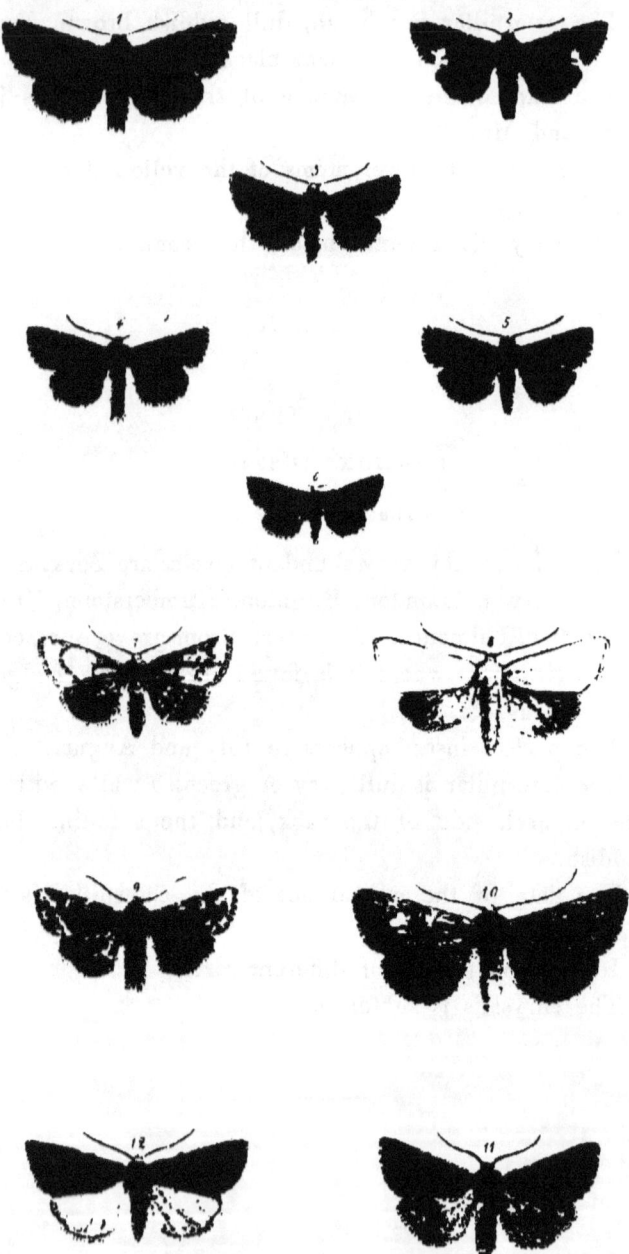

MIANA STRIGILIS.

MARKED MINOR.

Plate L. Figure 2.

Localities for this abundant species are York, Sutton-on-Derwent, Brighton, Falmouth, Darenth Wood, Preston, Black Park, Faversham, Carlisle, Bowdon, Glasgow, etc.

The situations where it is found are gardens, hedge-sides, woods, etc.

The perfect insect appears in June and July.

The caterpillar is greyish or pale greenish, lighter coloured beneath, the lines paler, the raised spots black.

The date of the appearance of the caterpillar is in April and May.

It feeds on the stems of grasses.

The chrysalis is found underneath the ground in a cocoon of earth.

MIANA FASCIUNCULA.

MIDDLE BARRED MINOR.

Plate L. Figure 3.

Localities for this species are York, Brighton, Falmouth, Canterbury, Faversham, Peterborough, Worcester, Lewes, Stowmarket, Birkenhead, Wavendon, Bristol, Kingsbury, Scarborough, Shrewsbury, Manchester, Burton-on-Trent, Darlington, Edinburgh, Halton, Huddersfield, Bowdon, Cambridge, Lower Guiting, etc.

The perfect insect appears at the end of June and in July.

The situations where it is found are hedge-sides, etc

MIANA LITEROSA.
ROSY MINOR.
Plate L. *Figure* 4.

Localities for this elegant species are York, Isle of Man, Brighton, Lewisham, Arundel, Tenterden, Exeter, Birkenhead, Cambridge, Deal, Stowmarket, Darlington, Edinburgh, Huddersfield, New Brighton, Scarborough, Lewes, Manchester, Bowdon, Lytham.

The situations where it is found are wood sides.

The perfect insect appears in June and July.

MIANA FURUNCULA.
CLOAKED MINOR.
Plate L. *Figure* 5.

Localities for this species are York, New Brighton, Sutton-on-Derwent, Brighton, Faversham, West Looe, Falmouth, Bowdon.

The perfect insect appears at the end of July.

MIANA EXPOLITA.
Plate L. *Figure* 6.

Localities for this species are near Galway, and Darlington.

The perfect insect appears in July; July 21.

The moth flies in the hot sunshine at times.

MIANA ARCUOSA.

LEAST MINOR.

Plate L. Figure 7.

Localities for this species are York, Stockton Forest, Sutton-on-Derwent, Darenth Wood, Whittlesea Mere, Lewes, Poynings, Brighton, Dulwich, Black Park, Torwood, Stirling, Birkenhead, Edinburgh, Bristol, Exeter, Shrewsbury, Lyndhurst, Manchester, Wavendon, Burton-on-Trent, Kingsbury, Worthing, Worcester, Bowdon.

The situations where it is found are wild grassy places.
The perfect insect appears in July.
It feeds in the stems of the hair-grass (*Aira cæspitosa*).
The chrysalis is found in an earthen cocoon.

I have to thank Mr. George Bunyard, of Maidstone, for the offer of valuable assistance.

MIANA BONDII.

Plate L. Figure 8.

Localities for this species are near Folkestone.
The situations where it is found are on the coast.
The perfect insect appears in July.

CELÆNA HAWORTHII.

HAWORTH'S MINOR.

Plate L. Figure 9.

Localities for this species are York, Askham Bog, Brighton, Simonswood Moss near Liverpool, Whittlesea Mere,

Windermere, Stirling, Darlington, Preston, Edinburgh, Manchester, Bowdon, Saddleworth, Arran, Scarborough, Kilmun.

The situations where it is found are heaths, and moist places.

The perfect insect appears in July, August, and September.

The date of the appearance of the caterpillar is in May and June.

It feeds on the cotton-grass (*Eriophorum angustifolium.*)

CARADRINIDÆ.

GRAMMESIA TRILINEA.

TREBLE-LINE.

Plate L. *Figure* 10.

Localities for this species are York, Lewes, Brighton, Epping, Birkenhead, Bristol, Hammersmith, Burton-on-Trent, Huddersfield, Scarborough, Lewisham, Cambridge, Kingsbury, Manchester, Shrewsbury, Black Park, Exeter, Lyndhurst, Stowmarket, Tenterden, Peterborough, Halton, Wavendon, Worthing, Worcester, Bowdon.

The situations where it is found are gardens, woods, hedge-sides, etc.

The perfect insect appears at the end of May, and in June and July.

The caterpillar is reddish brown, with a whitish line along the back, and another one of dark brown below it; the side-line dark brown, and on each side a row of black dots.

The date of the appearance of the caterpillar is in April and May.

It feeds on the plantain (*Platago lanceolata*).

The chrysalis is found below the surface of the earth.

A permanent variety of this species has the central shade darker.

I have to thank Mr. Thomas Blackburn, of Bowdon, Cheshire, for extensive information about the insects of that district.

HYDRILLA PALUSTRIS.

Plate L. *Figure* 11.

Localities for this species are at Crompton's Coppice near York.

The situations where it is found are moist places in woods.

The perfect insect appears at the end of May, and in June and July.

The caterpillar is brownish, with a whitish line along the back, and two rows of white dots on each side, and raised spots of black.

The date of the appearance of the caterpillar is in July and August.

It feeds on the plantain (*Plantago lanceolata*), and other low plants.

The chrysalis is subterranean.

ACOSMETIA CALIGINOSA.

REDDISH BUFF.

Plate L. *Figure* 12.

Localities for this species are Lyndhurst in the New Forest, and near Ryde in the Isle of Wight.

The situations where it is found are moist places in woods.

The perfect insect appears in June and July.

This moth flies in the day-time as will as at dusk.

CARADRINA MORPHEUS.
BORDERED RUSTIC.
Plate LI. *Figure* 1.

Localities for this plentiful species are York, Sutton-on-Derwent, Scarborough, Brighton, Humberstone, Exeter, Falmouth, Edinburgh, Faversham, Cambridge, Bristol, Shrewsbury, Worthing, Black Park, Burton-on-Trent, Lewes, Wavendon, Birkenhead, Manchester, Stowmarket, Worcester, Bowdon.

The situations where it is found are woods, gardens, hedge-sides, etc.

The perfect insect appears in July.

The caterpillar is greyish brown, with a row of black wedge-shaped streaks on each side of the back; the side line paler.

The date of the appearance of the caterpillar is in September, and on to April.

It feeds on the teazle and other plants.

The chrysalis is found beneath the ground.

CARADRINA ALSINES.
DOTTED RUSTIC.
Plate LI. *Figure* 2.

Localities for this also common species are York, Wallasey near Birkenhead, Brighton, etc., etc., etc.

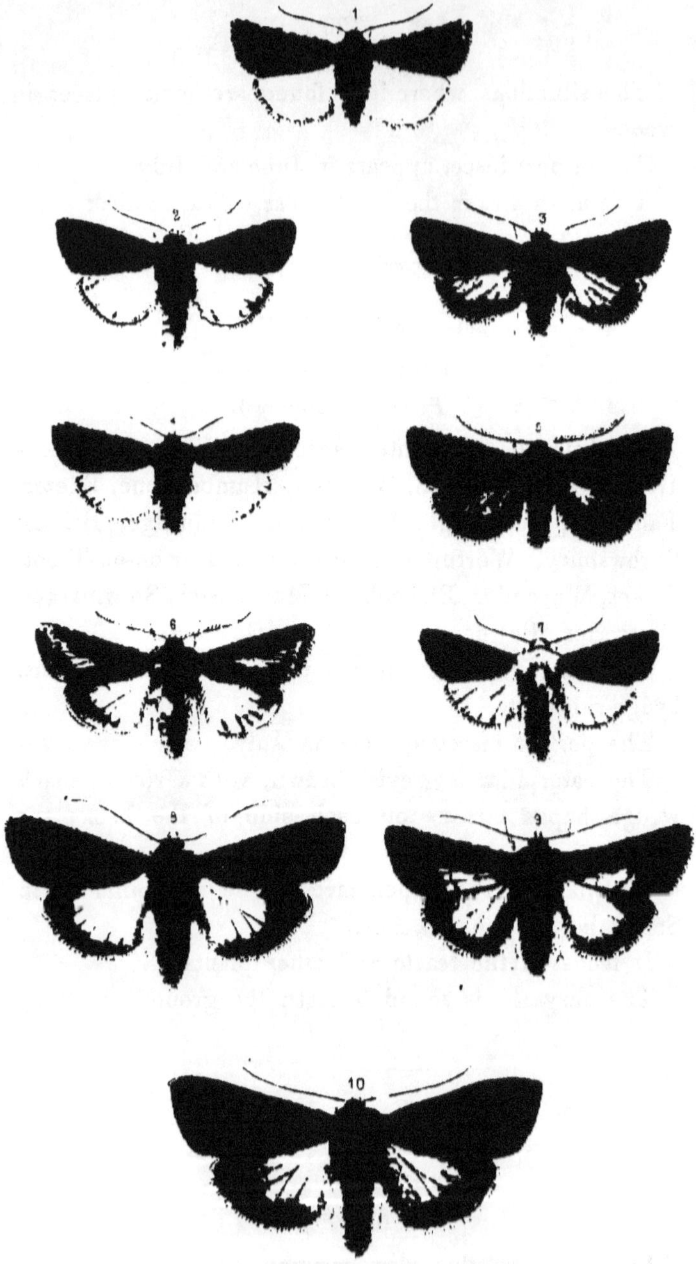

The situations where it is found are gardens, hedge-sides, etc.

The perfect insect appears in July.

The caterpillar is greyish, streaked on the sides with a pale line on each side of the back bordered above by another fine one; the side line also pale and broad, with some slanting blackish spots above it.

The date of the appearance of the caterpillar is in March and April.

It feeds on the dock (*Rumex pratensis*), the chickweed *Alsine media*), and the plantain (*Plantago lanceolata*).

The chrysalis is subterranean.

CARADRINA BLANDA.

Plate LI. *Figure* 3.

Localities for this species are York, Scarborough, Brighton, Birkenhead, Edinburgh, Lewisham, Bristol, Exeter, Manchester, Faversham, Cambridge, Wavendon, Halton, Tenterden, Black Park, Lower Guiting, Lewes, Kingsbury, Uppingham, Darlington, Worthing.

The situations where it is found are woods, etc.

The perfect insect appears in June and July.

The caterpillar is greyish, streaked on the sides, with a pale line on each side of the back, bordered above by another fine slanting one, the side line also pale and broad, with some slanting blackish spots above it.

It feeds on various low plants.

The chrysalis is found underneath the surface.

CARADRINA CUBICULARIS.

PALE MOTTLED WILLOW.

Plate LI. *Figure* 4.

Localities for this very common species are York, Sutton-on-Derwent, Brighton, Falmouth, Birkenhead, Bowdon, etc.

The perfect insect appears in June, July, and August.

The caterpillar is reddish grey on the back, the sides dark grey; the head shining black.

The date of the appearance of the caterpillar is in September and through the winter on to April.

It feeds on the chickweed (*Alsine media*).

The chrysalis is subterranean in a cocoon.

This species comes much to a light.

NOCTUIDÆ.

RUSINA TENEBROSA.

BROWN FEATHERED RUSTIC.

Plate LI. *Figure* 5.

Localities for this species are York, Beverley, Stockton Forest, Doncaster, Darenth Wood, Stirling, Brighton, Birkenhead, Bristol, Wimbledon Common, Faversham, Edinburgh, Burton-on-Trent, Canterbury, Stowe Wood, Sherwood Forest, Darlington, Exeter, Black Park, Lewes, Huddersfield, Stowmarket, Lyndhurst, Dorking, Reigate, Kingsbury, Manchester, Worthing, Bere Forest, Lynton, Bowdon, Tenterden.

The situations where it is found are woods.

The perfect insect appears in June and July.

The caterpillar is rich brown, with a dark line along the back, the first five segments having a slender white line through the centre, a line composed of black streaks on each side below it, and bordered above on its hinder portion with pale brown; the side line pale brown.

The date of the appearance of the caterpillar is in September and on to March.

It feeds on the knot grass (*Polygonum aviculare*), and other low plants.

The chrysalis is found in a cocoon a little below the surface.

AGROTIS VALLIGERA.

ARCHED DART.

Plate LI. Figure 6.

Localities for this species are York, Sutton-on-Derwent, New Brighton, Isle of Man, Solway Frith, Tremori, Birkenhead, Darlington, Edinburgh, Stowmarket, Deal, Belcurrick.

The situations where it is found are sand hills on the coast.

The perfect insect appears in July and August.

The caterpillar is dull greenish grey, with a paler line along the back and another on either side of it, and two rows of black dots between them; also a row of short white streaks on the lower part of the sides.

The date of the appearance of the caterpillar is in October and thence to May.

It feeds on the roots of grasses.

The chrysalis is subterranean.

AGROTIS PUTA.

SHUTTLE-SHAPED DART.

Plate LI. *Figure* 7.

Localities for this species are Faversham, Hammersmith, Sutton-on-Derwent, Brighton, Birkenhead, Bristol, Lewes, Plymouth, Stowmarket, Lewisham, Cambridge, Halton, Lyndhurst, Tenterden, Arundel, Exeter, Peterborough, Wavendon, Newhaven, and near London.

The situations where it is found are woods, gardens, etc.
The perfect insect appears in August and September.
The chrysalis is found beneath the surface.

AGROTIS SUFFUSA.

DARK SWORD-GRASS.

Plate LI. *Figure* 8.

Localities for this now common but formerly considered rare species are Liverpool, Falmouth, Sutton-on-Derwent, Nunburnholme, Crompton's Coppice near York, Darenth Wood, Birch Wood, Epping, Brighton, Dulwich, New Forest, Chichester, Plymouth, Hammersmith, Barnstaple, Peterborough, Chester, Black Park, Huddersfield, Carron, Glasgow, Kilmun, Faversham, Bowdon, Stirling, Rannock, and Boyd's Planting near Torwood.

The situations where it is found are woods, gardens, etc.
The perfect insect appears in September and on to February; February 3, 28; also in June (?)
The caterpillar is shining grey.

The date of the appearance of the caterpillar is in May.
It feeds on the roots of grasses.
The chrysalis is found beneath the ground.

AGROTIS FENNICA.

Plate LI. Figure 9.

Localities for this species are in Derbyshire—the exact locality I at present am unacquainted with.

The perfect insect appears in July and August.

AGROTIS SAUCIA.

PEARLY UNDERWING.

Plate LI. Figure 10.

Localities for this species are York, Wallasey near Birkenhead, Brighton, Shrewsbury, Hammersmith, Bristol, Killarney, Cambridge, Worcester, Lewisham, Lewes, Arundel, Lyndhurst, Deal, New Forest, Manchester, Plymouth, Warrington, Barnstaple, Ventnor, Bonchurch, Isle of Wight, and near London.

The situations where it is found are where the ivy blooms.

The perfect insect appears in July, August, and September; September 28.

The caterpillar is greyish brown, with a paler line along the back running through a row of angular-shaped darker spots; the side line and side spots dark brown.

The date of the appearance of the caterpillar is in November.

It feeds on the plantain (*Plantago lanceolata*), and the dock (*Rumex pratensis*).

The chrysalis is found underneath the ground.

AGROSTIS SEGETUM.

COMMON DART.

Plate LII. *Figure* 1.

Localities for this most abundant species are York, Sutton-on-Derwent, Brighton, Humberstone, Falmouth, Charmouth, Nunburnholme, Bromsgrove, Worcester, Nafferton, Bowdon, etc., etc., etc., etc.

The situations where it is found are gardens, hedgesides, etc.

The perfect insect appears in June, July, and September.

The caterpillar is greenish grey with a paler line along the back, and a pale brown one on each side of it; the spots black.

The date of the appearance of the caterpillar is from September to May.

It feeds on the roots of various grasses.

The chrysalis is found beneath the earth.

AGROTIS LUNIGERA.

Plate LII. *Figure* 2.

Localities for this species are Edinburgh, Liverpool, Cork, Ventnor and Bonchurch in the Isle of Wight.

The perfect insect appears in June, July, and August.

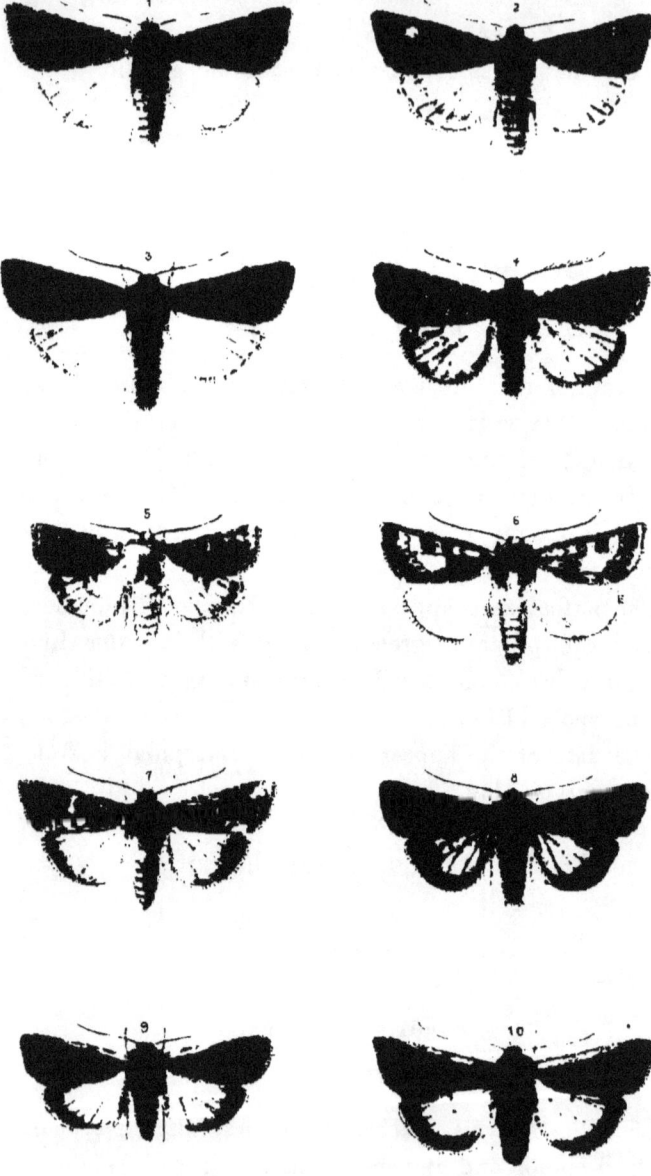

AGROTIS EXCLAMATIONIS.

HEART AND DART.

Plate LII. *Figure* 3.

Localities for this very abundant species are York, Sutton-on-Derwent, Nunburnholme, Charmouth, Bowdon, Worcester, Bromsgrove, Brighton, Humberstone, Falmouth, etc., etc.

The situations where it is found are gardens, hedgesides, etc.

The perfect insect appears in June, July, and August.

The caterpillar is dull grey with a paler line on each side of the back; side line the same; the spots black.

The date of the appearance of the caterpillar is in September and on to May.

It feeds on the roots of grasses and other low plants.

The chrysalis is found beneath the surface.

AGROTIS CORTICEA.

HEART AND CLUB.

Plate LII. *Figure* 4.

Localities for this species are York, Brighton, Chester, Falmouth, Lewisham, Ventnor, Bonchurch, Birkenhead, Hammersmith, Edinburgh, Lewes, Clapham, Cambridge, Bristol, Halton, Plymouth, Black Park, Stowmarket, Worcester, Arundel, Tenterden, Wavendon.

The situations where it is found are woods, gardens, etc.

The perfect insect appears in July.

AGROTIS CINEREA.

LIGHT FEATHERED RUSTIC.

Plate LII. *Figure* 5.

Localities for this species are Carlisle, Brighton, Sevenoaks, Llanferrar, Bristol, Halton, Lewes.

The perfect insect appears in June.

The caterpillar is shining greenish brown, with a darker line along the back and another on each side of it, and between them some small slanting dark streaks.

The date of the appearance of the caterpillar is September and to May.

It feeds on the roots of various low plants.

The chrysalis is subterranean.

AGROTIS RIPÆ.

Plate LII. *Figure* 6.

Localities for this species are Runcorn, Barnstaple.

The situations where it is found are sand-hills on the coast.

The perfect insect appears in June and July.

AGROTIS CURSORIA.

Plate LII. *Figure* 7.

Localities for this species are Bristol, Yarmouth, New Brighton, Lytham, Solway Firth, Birkenhead, Edinburgh.

The situations where it is found are sand-hills on the coast.

The perfect insect appears in July and August.

The caterpillar is pale dull yellowish, with a dark brown line along the back; the side line whitish edged on its upper part with brown; spots black.

The date of the appearance of the caterpillar is in May and June.

It feeds on the spurge *(Euphorbia esula)*.

The chrysalis is found under the ground.

AGROTIS NIGRICANS.

GARDEN DART.

Plate LII. *Figure* 8.

Localities for this species are York, Sutton-on-Derwent, Stockton Forest, Brighton, Darlington, Halton, Arundel, Bristol, Edinburgh, Huddersfield, Black Park, Stirling, Cambridge, Exeter, Kingsbury, Carron, Scarborough, Manchester, Shrewsbury, Stowmarket, Bowdon.

The situations where it is found are woods.

The perfect insect appears in June, July, and August.

The caterpillar is shining brown dotted with black, and with an indented line of a paler shade.

The date of the appearance of the caterpillar is in May and June.

It feeds on different low plants.

The chrysalis is found below the surface.

AGROTIS TRITICI.

WHITE LINED DART.

Plate LII. *Figure* 9.

Localities for this species are York, Brighton, Lewes, Falmouth, Deal, Plymouth, New Brighton, Isle of Man, Solway Frith, Birkenhead, Exeter, Bristol, Manchester, Darlington, Scarborough, Edinburgh, Stowmarket, Bowdon.

The perfect insect appears in August.

The caterpillar in shining grey.

The date of the appearance of the caterpillar is in May.

It feeds on grass and other low plants.

The chrysalis is subterranean.

AGROTIS AQUILINA.

Plate LII. *Figure* 10.

Localities for this species are York, Stockton Forest, Scarborough, Brighton, Barnstaple, Yarmouth, Plymouth, Birkenhead, Lewes, New Brighton, Cambridge, Bowdon, Manchester, and near London.

The perfect insect appears in July and August.

The caterpillar is very pale brown, the sides brownish grey, with black dots; the head pale brown.

The date of the appearance of the caterpillar is in May.

It feeds on the yellow ladies' bedstraw (*Galium verum*).

The chrysalis is subterranean.

AGROTIS OBELISCA.

Plate LIII. *Figure* 1.

Localities for this species are Sutton-on-Derwent, Scarborough, Brighton, Ventnor, Bonchurch, Isle of Wight, Dublin, Edinburgh.

The perfect insect appears in August and September.

The caterpillar is pale brown, with a dark grey stripe along the back, the black dots hardly perceptible; the head rather darker.

The date of the appearance of the caterpillar is in May.

It feeds on different low plants.

The chrysalis is found below the earth.

AGROTIS AGATHINA.

Plate LIII. *Figure* 2.

Localities for this species are York, Stockton Forest, Simonswood Moss near Liverpool, Black Park, Weybridge, Lyndhurst, Ticehurst, West Wickham.

The situations where it is found are heaths and heathy places in woods.

The perfect insect appears in August.

It feeds on heath.

The chrysalis is found underneath the ground.

AGROTIS PORPHYREA.

TRUE-LOVER'S KNOT.

Plate LIII. *Figure* 3.

Localities for this species are York, Langwith, Allerthorpe Common, Stockton Forest, Lewes, Black Park, Birkenhead, Lynton, Bristol, Exeter, Rudheath, Edinburgh, Huddersfield, Lyndhurst, Manchester, Tertorden, Scarborough, Shrewsbury, Wavendon, Leeds, Reigate.

The situations where it is found are heaths.

The perfect insect appears in June and July.

The caterpillar is reddish orange, with a row of conspicuous white spots along the back, edged with blackish; the head brown.

The date of the appearance of the caterpillar is in August.

It feeds on heath.

The chrysalis is found underneath the ground.

AGROTIS PRÆCOX.

DUCHESS OF PORTLAND.

Plate LIII. *Figure* 4.

Localities for this rather rare species are Nunburnholme, Portland, Lytham, Birkenhead, Edinburgh, Manchester, New Brighton, Shrewsbury.

The situations where it is found are gardens, and on the sea-coast.

The perfect insect appears in August.

The caterpillar is dull yellowish with a whitish line along the back; and the second, twelfth, and thirteenth segments whitish; the side line also whitish.

The date of the appearance of the caterpillar is in May. It feeds on various low plants.

The chrysalis is subterranean.

AGROTIS RAVIDA.

STOUT DART.

Plate LIII. *Figure* 5.

Localities for this species are York, Scarborough, Nun-.burnholme, Nafferton, Osgodby, Anstey, and near London, Monkswood, Worcester, Lewisham, Stowmarket, Halton, Edinburgh, Darlington, Cambridge, Burton-on-Trent.

The situations where it is found are gardens.

The perfect insect appears in June and July.

The caterpillar is pale brownish dull yellow with a paler line along the back, and a white one on each side of it, bordered above with a row of white dots; the side line whitish, with a row of brownish spots above it.

The date of the appearance of the caterpillar is in April.

It feeds on the dock (*Rumex pratensis*), and other low plants.

The chrysalis is found below the earth.

AGROTIS PYROPHILA.

Plate LIII. *Figure* 6.

Localities for this species are York, Nunburnholme,

Bidston near Birkenhead, Old Swan near Liverpool, Edinburgh, Worcester, Wavendon, Halton.

The situations where it is found are gardens, etc.

The perfect insect appears in June and July.

The caterpillar is dull greyish brown.

The date of the appearance of the caterpillar is in April.

It feeds on grasses and low plants.

The chrysalis is found underneath the surface of the earth.

AGROTIS LUCERNA.

Plate LIII. *Figure* 7.

Localities for this neat species are near Forfar, Dover, Folkstone, Swansea, Llanferras, Arthur's Seat near Edinburgh, Keswick, Plymouth, Rannock.

The situations where it is found are the sides of mountains.

The perfect insect appears in July.

The caterpillar is dark greenish grey, with a double row of yellowish white spots, each of which is bordered on its front side by a black shade.

The date of the appearance of the caterpillar is in February and March.

It feeds on the dandelion (*Taraxicum dens-leonis*) and other low plants.

The chrysalis is found beneath the surface.

AGROTIS ASHWORTHII.

Plate LIII. *Figure* 8.

Localities for this species are Llanferras, Llangollen,

Clwyd, Llanrwst, Snowdon, and other places in North Wales.

The perfect insect appears in July.

The caterpillar is green, with a paler line along the back; the side line whitish, over it being a row of black spots.

It feeds on the heath, also on the hare-bell, (*Campanula rotundifolia*.)

The chrysalis is found under the earth.

This species flies briskly by day.

TRIPHÆNA JANTHINA.

LESSER BROAD-BORDERED

Plate LIII. *Figure* 9.

Localities for this species are York, Sutton-on-Derwent, Charmouth, Brighton, Bowdon, Faversham, Isle of Wight, Barnstaple, Peterborough.

The situations where it is found are hedge-sides, gardens, and woods.

The perfect insect appears in July and August.

The caterpillar is dull greyish yellow with a paler line along the back, and a pair of conspicuous black spots on the ninth, tenth, eleventh, and twelfth segments; the side line spots white.

The date of the appearance of the caterpillar is in April.

It feeds on the primrose and other low plants.

The chrysalis is found beneath the ground.

This species flies early in the evening.

TRIPHÆNA FIMBRIA.

BROAD-BORDERED YELLOW UNDERWING.

Plate LIV. *Figure* 1.

Localities for this species, formerly thought so extremely rare, (my friend, Mr. Dale, of Glanville Wootton, had, for several years, with a few other specimens, the wing of one in his cabinet which he had found in a spider's web,) but which is now taken plentifully, are York, Langwith, Stockton Forest, Sandal Beat, and Melton Wood near Doncaster, Bolton Abbey, Sutton-on-Derwent, Brighton, Faversham, Canterbury, Lytham, Swinhope, Ticehurst, Arundel, Horndean, Carlisle, Lewisham, Milstead near Sittingbourne, Ilfracombe, Sherwood Forest, Killarney, West Wickham, Bristol, Bowdon, Black Park, Dorking, Birkenhead, Edinburgh, Darlington, Burton-on-Trent, Exeter, Halton, Lewes, Huddersfield, Lyndhurst, Pembury, Plymouth, Manchester, Scarborough, Stowmarket, Tenterden, Shrewsbury, Worthing, Worcester.

The perfect insect appears in June and July.

The caterpillar is brown with a paler line along the back, and a row of slanting whitish stripes, alternating with some white spots in the place of the usual line on either side of the back.

The date of the appearance of the caterpillar is in March and April.

It feeds on the primrose and other low plants.

The chrysalis is found beneath the surface.

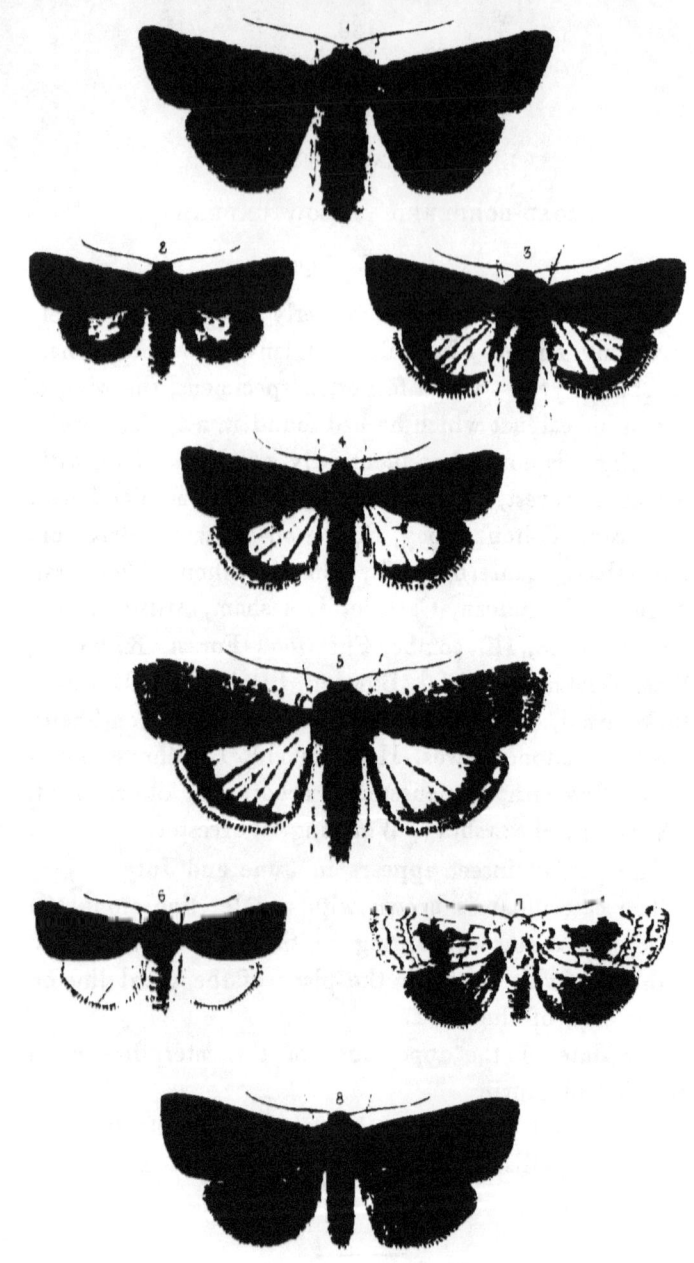

TRIPHÆNA INTERJECTA.

LEAST BROAD-BORDERED YELLOW UNDERWING.

Plate LIV. *Figure 2.*

Localities for this species are York, Sutton-on-Derwent, Langwith, Charmouth, Lewisham, Birkenhead, Brighton, Halton, Newhaven, Bristol, Cambridge, Kingsbury, Manchester, Plymouth, Scarborough, Isle of Wight, Lower Guiting, Lyndhurst, Pembury, Shrewsbury, Stowmarket, Lewes, Worthing, Peterborough, Burton-on-Trent, Tenterden, Worcester, Wavendon, and near London.

The situations where it is found are hedge-sides, gardens, etc.

The perfect insect appears in July and August.

The caterpillar is dull whitish yellow, streaked with pale yellow and brownish, the line along the back white and narrow, the line on each side below it white, edged above and below with reddish; the side line pale, edged on its upper side with a broad brownish stripe.

The date of the appearance of the caterpillar is in March and April.

It feeds on different low plants.

The chrysalis is subterranean.

This species flies early in the evening.

TRIPHÆNA SUBSEQUA.

Plate LIV. *Figure 3.*

Localities for this species, which is distinguished from *Triphæna orbona*, by a black spot or mark on the upper margin of the fore wings near the tip, are the Isle of Bute, Lyndhurst, New Forest, Sherwood Forest.

The situations where it is found are heaths.

The perfect insect appears in July; July 27.

The caterpillar is brownish grey with a slight tinge of greenish, and a paler line along the back and another below it on each side, the latter having a row of blackish square spots above it; the second segment dark brown.

The date of the appearance of the caterpillar is in March and April.

It feeds on different low plants.

The chrysalis is found beneath the surface.

TRIPHÆNA ORBONA.

LESSER YELLOW UNDERWING.

Plate LIV. *Figure* 4.

Localities for this common species are Nunburnholme, York, Sutton-on-Derwent, Askham Bog, Isle of Wight, Charmouth, Anstey, Brighton, Falmouth, Bowdon, Black Park, Faversham, Barnstaple, and near London.

The situations where it is found are gardens, woods, lanes, and hedgerows.

The perfect insect appears in August and September; August 21, 23; September 16.

The caterpillar is dull yellowish brown, mottled with darker on the back, and with a conspicuous side-line of a paler shade; the side spots inserted in large brownish blots.

The date of the appearance of the caterpillar is in March and April.

It feeds on various low plants.

The chrysalis is found underneath the surface of the ground.

TRIPHÆNA PRONUBA.

YELLOW UNDERWING.

Plate LIV. *Figure* 5.

Localities for this most abundant species are York, Sutton-on-Derwent, Askham Bog, Nunburnholme, Anstey, Nafferton, Brighton, Charmouth, Falmouth, Birkenhead, Worcester, Bowdon, Bromsgrove, and near London.

The situations where it is found are gardens, woods, lanes, and hedgerows.

The perfect insect appears in June, July, and August.

The caterpillar is dull greyish or yellowish green, with a paler line along the back, and another similar one on the sides below it, the latter with some large blackish spots above it on the third, fourth, fifth, sixth, seventh, eighth, ninth, tenth, and eleventh segments; the side line indistinct.

The date of the appearance of the caterpillar is in April.

It feeds on almost all low plants, the primrose, etc., etc.

The chrysalis is found beneath the earth.

NOCTUA GLAREOSA.

AUTUMNAL RUSTIC.

Plate LIV. *Figure* 6.

Localities for this species are York, Buttercrambe Moor, Stockton Forest, Scarborough, Huddersfield, Brighton, Isle of Man, Storeton near Birkenhead, Wootton near Liverpool, Birch Wood, the New Forest, Edinburgh, Dorking, Weybridge, Lewes, Chichester, Lyndhurst, Manchester, Sherwood Forest, Plymouth, Rotherham, Tenterden, and Stirling.

The situations where it is found are heaths and heathy places in woods, etc.

The perfect insect appears in August.

The caterpillar is pale brown, with a paler line along the back, and another like one below it on each side, edged with dark brown; the side line yellowish white.

The date of the appearance of the caterpillar is in June.

The chrysalis is found below the ground.

NOCTUA DEPUNCTA.

Plate LIV. *Figure* 7.

Localities for this species are Doncaster, Scarborough, Sutton-on-Derwent, Carlisle, Exeter, Manchester, Stowmarket.

The situations where it is found are woods.

The perfect insect appears in July and August.

The caterpillar is greyish brown, with a whitish line on either side of the back, with a row of black dots; the dots on the side white, in black rings.

It feeds on the sorrel and other low plants.

The chrysalis is found underneath the surface.

NOCTUA AUGUR.

DOUBLE DART.

Plate LIV. *Figure* 8.

Localities for this sufficiently common species are York,

Sutton-on-Derwent, Anstey, Brighton, Bowdon, etc., etc.

The situations where it is found are gardens, hedge-sides, etc.

The perfect insect appears in June and July.

The caterpillar is brownish orange, with a paler line along the back, and one below it on either side formed of slanting black streaks alternated with white spots, the spots on the side white, edged on each side with yellowish; the head dark brown.

The date of the appearance of the caterpillar is in April and May.

It feeds on various low plants.

The chrysalis is found below the earth.

NOCTUA PLECTA.

FLAME-SHOULDER.

Plate LV. *Figure* 1.

Localities for this also common species are York, Sutton-on-Derwent, Brighton, Falmouth, Bowdon, etc., etc.

The perfect insect appears in June and August?

The caterpillar is dull reddish brown, with a line on each side of the back, composed of a row of white dots.

The date of the appearance of the caterpillar is in April.

It feeds on different low plants.

The chrysalis is subterranean.

NOCTUA FLAMMATRA.

Plate LV. *Figure* 2.

Localities for this species are the Isle of Wight.

The caterpillar is green with paler stripes on the sides.
The date of the appearance of the caterpillar is in April.
It feeds on several low plants.
The chrysalis is found below the earth.

NOCTUA C-NIGRUM.

SETACEOUS HEBREW CHARACTER.

Plate LV. *Figure* 3.

Localities for this, which is another common species, are York, Sutton-on-Derwent, Nunburnholme, Bromsgrove, Charmouth, Lewisham, Plumstead, Black Park, Brighton, Chichester, Bere Forest, Bowdon, Sidmouth, Plymouth, Crewe, Peterborough, Stirling.

The situations where it is found are gardens, and the sides of water courses and brooks.

The perfect insect appears in July and August. July 8.

The caterpillar is greenish grey, with a yellowish line along the back, a pale green one on either side below it, and a white side-line, the space between these two last-named being dark green.

The date of the appearance of the caterpillar is in April.
It feeds on several low plants.
The chrysalis is subterranean.

NOCTUA DITRAPEZIUM.

Plate LV. *Figure* 4.

Localities for this species are Darenth Wood, Ripley near London, Epping, Blandford, Galway, Leith Hill in Surrey, Portsmouth, Birch Wood, and the New Forest.

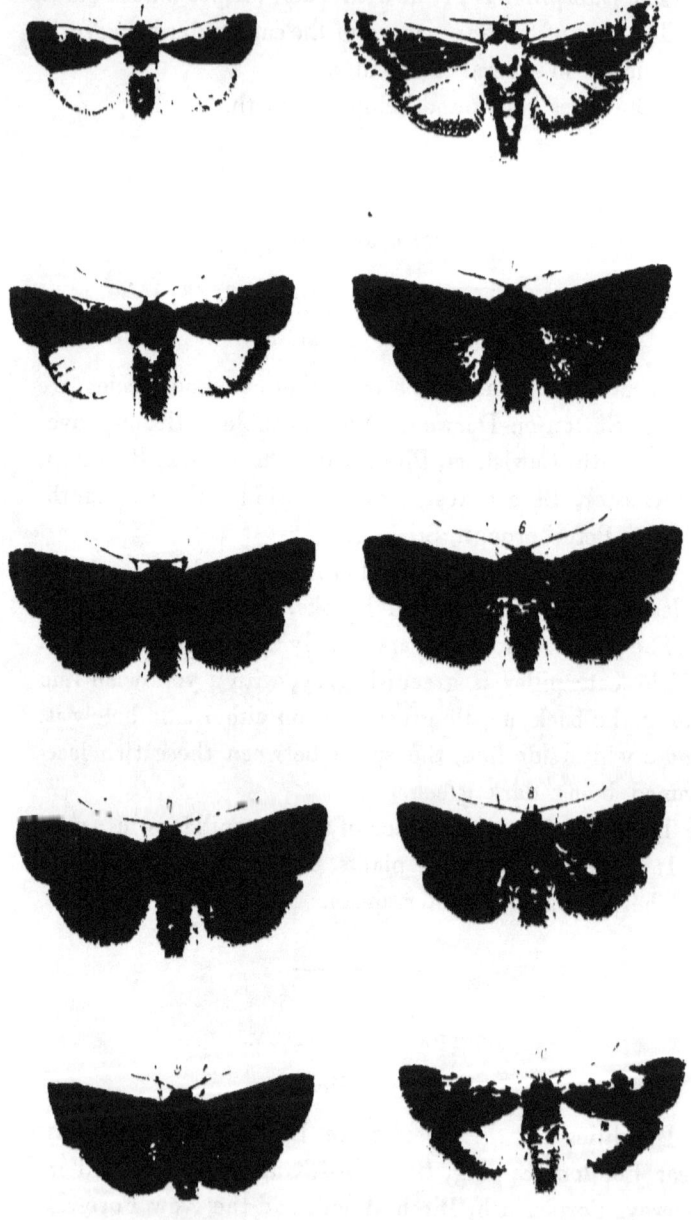

The situations where it is found are woods.

The perfect insect appears in July.

The caterpillar is dull greyish yellow slightly tinged with red, with darker marks along the back, most conspicuous on the eleventh and twelfth segments.

The date of the appearance of the caterpillar is in April.

It feeds on various low plants.

The chrysalis is found below the ground.

NOCTUA TRIANGULUM.

DOUBLE SQUARE-SPOT.

Plate LV. *Figure* 5.

Localities for this species are York, Burnby, Sutton-on-Derwent, Stockton Forest, Brighton, Black Park, Lewes, Bowdon, Dorking, Birkenhead, Exeter, Arundel, Bristol, Kingsbury, Lyndhurst, Bere Forest, Burton-on-Trent, Manchester, Sudbury, Cambridge, Plymouth, Tenterden, Preston, Darlington, Stowmarket, Worthing.

The situations where it is found are woods.

The perfect insect appears in June.

The caterpillar is dull reddish yellowish, mottled with darker along the back, and with some elongated black spots along the line below the back line, on the eighth, tenth, eleventh, and twelfth segments, most conspicuous on the two last-named.

The date of the appearance of the caterpillar is in April.

It feeds on various low plants.

The chrysalis is found below the ground.

NOCTUA RHOMBOIDEA.

Plate LV. *Figure* 6.

Localities for this species are Scarborough, Linacre near Liverpool, Darenth Wood, Ripley near London, Epping, Bristol, Black Park, Darlington, Reigate, Wavendon, Arundel, Worthing, Chesterfield.

The situations where it is found are woods.

The perfect insect appears in July.

The caterpillar is reddish brown, with a whitish line along the back, and another of the same on either side of it, the latter interrupted on each segment, but connected above with whitish spots.

The date of the appearance of the caterpillar is in April and May.

It feeds on various low plants.

The chrysalis is found below the earth.

NOCTUA BRUNNEA.

PURPLE CLAY.

Plate LV. *Figure* 7.

Localities for this species are York, Sutton-on-Derwent, Stockton Forest, Brighton, Perth, Bowdon, Bristol, Bere Forest, Darenth Wood, Birkenhead, Burton-on-Trent, Cambridge, Kingsbury, Scarborough, Preston, Islington, Lewes, Manchester, Stowmarket, Isle of Man, Lyndhurst, Huddersfield, Tenterden, Wavendon, Worthing.

The situations where it is found are woods.

The perfect insect appears in July.

The caterpillar is brown, mottled with yellow along the back, which has a series of slanting black spots, on which are some conspicuous yellow spots; the side line greyish yellow.

It feeds on different low plants.

The chrysalis is subterranean.

NOCTUA FESTIVA.

ENGRAILED CLAY.

Plate LV. *Figure* 8.

Localities for this species are York, Nunburnholme, Sutton-on-Derwent, Stockton Forest, Charmouth, Bowdon, Brighton, Falmouth, Darenth Wood, Bere Forest, Barnstaple, Torwood, Stirling.

The situations where it is found are woods, hedgesides, etc.

The perfect insect appears in June and July.

The caterpillar is reddish, with a slight mixture of dull yellowish or grey, darker along the back, and with a row of oblong blackish spots on each side below the back, on the fifth, sixth, seventh, eighth, ninth, tenth, eleventh, and twelfth segments.

The date of the appearance of the caterpillar is in April and May.

It feeds on various low plants.

The chrysalis is found underneath the surface.

NOCTUA DAHLII.

BARRED CHESNUT.

Plate LV. *Figure* 9.

Localities for this species are Hargrave Hall and Hooton near Birkenhead, Lewes, Bowdon, Carlisle, Arran, West Wickham, Bristol, Black Park, Edinburgh, Lewes, Horndean, Huddersfield, Lyndhurst, Plymouth, Sherwood Forest, Manchester, Pembury.

The situations where it is found are woods, etc.

The perfect insect appears in July and August.

The caterpillar is reddish mixed with grey, with a paler line along the back, and another similar one on each side below it, with a row of black dots in white rings above the latter.

NOCTUA SUBROSEA.

Plate LV. *Figure* 10.

Localities for this species are Whittlesea Mere and Yaxley Fen, the Fens of the counties of Cambridge and Huntingdon.

The situations where it is found are the Fens.

The perfect insect appears in July.

The caterpillar is dull reddish yellow marbled with brown, with a yellow line edged with brown along the back, and another of the same colours on either side below it; the side line yellow and broad.

The date of the appearance of the caterpillar is in May and June.

It feeds on the bog myrtle *(Myrica gale)*.

The chrysalis is found below the earth.

NOCTUA RUBI.

SMALL SQUARE-SPOT.

Plate LVI. *Figure* 1.

Localities for this species are York, Sutton-on-Derwent, Stockton Forest, Manchester, Lewes, Black Park, Uppingham, Rudheath, Preston, Birkenhead, Bristol, Exeter, Edinburgh, Bowdon, Burton-on-Trent, Halton, Cambridge, Kingsbury, Lyndhurst, Shrewsbury, Worthing, Scarborough, Wavendon.

The situations where it is found are woods, etc.

The perfect insect appears in June and in August.

The caterpillar is greenish grey, with a darker-edged white line along the back, and that on the sides beneath it with obscure slanting stripes across; the side line pale dull greenish yellow, edged above with a darker shade.

The date of the appearance of the caterpillar is in June and July.

It feeds on various low plants.

The chrysalis is subterranean.

NOCTUA UMBROSA.

SIX STRIPED RUSTIC.

Plate LVI. *Figure* 2.

Localities for this species are York, Nunburnholme, Sutton-on-Derwent, Stockton Forest, Brighton, Lewisham, Bowdon, Shrewsbury, Manchester, Scarborough, Bristol,

Wavendon, Deptford, Birkenhead, Worthing, Barnstaple, Burton-on-Trent, Darlington, Peterborough, Edinburgh, Halton, Huddersfield, Lewes, New Brighton, Kingsbury, Exeter, Lyndhurst.

The situations where it is found are gardens, woods, etc.

The perfect insect appears in August.

The caterpillar is whitish grey, with a black line on each side of the back.

The date of the appearance of the caterpillar is in April and May.

It feeds on grasses and other ground plants.

The chrysalis is found below the earth.

NOCTUA BAJA.

DOTTED CLAY.

Plate LVI. Figure 3.

Localities for this rather common species are York, Sutton-on-Derwent, Brighton, Burnby, Bowdon, etc., etc., etc.

The situations where it is found are woods and gardens.

The perfect insect appears in July.

The caterpillar is dull yellowish, marbled with brown, with a yellowish line edged with black along the back, and another of the like colour on either side below it, from which proceeds, on the fifth, seventh, eighth, ninth, tenth, eleventh, and twelfth segments, a slanting yellow streak.

The date of the appearance of the caterpillar is in April and May.

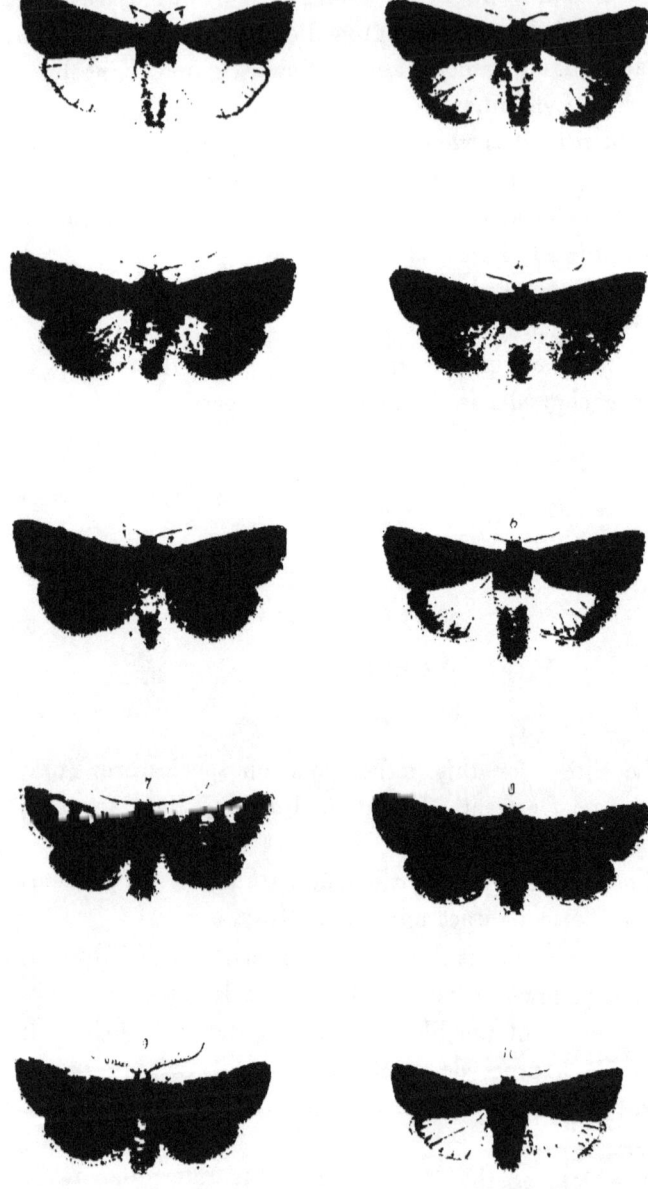

It feeds on various low plants.

The chrysalis is found under the ground.

I beg leave here to thank the Rev. Edward Horton, of Wick, near Worcester, an "old Bromsgrovian" schoolfellow, for useful information.

NOCTUA SOBRINA.

Plate LVI. *Figure* 4.

Localities for this species are at Rannock, in Perthshire. The perfect insect appears in July.

The caterpillar is blue grey, marbled with yellowish white, with a dull yellowish line along the back, and another interrupted one of the same colour below it on the sides; the side line pale greyish, edged on its upper margin with darker.

The chrysalis is subterranean.

NOCTUA NEGLECTA.

NEGLECTED RUSTIC.

Plate LVI. *Figure* 5.

Localities for this species are York, Stockton Forest, Hale near Liverpool, The New Forest, Birch Wood, West Wickham Wood, Lyndhurst, Pembury, Weybridge, Black Park, Ticehurst, Plymouth, and Carlisle.

The situations where it is found are heaths and heathy places in woods.

The perfect insect appears in August.

The caterpillar is at first green, but afterwards becomes dull reddish, or brownish.

The date of the appearance of the caterpillar is in October and on to May.

It feeds on heather at first, and afterwards on different low plants.

The chrysalis is found below the surface.

NOCTUA XANTHOGRAPHA.

SQUARE-SPOT RUSTIC.

Plate LVI. *Figure* 6.

Localities for this excessively common and very variable species are York, Nunburnholme, Sutton-on-Derwent, Charmouth, Brighton, Falmouth, Faversham, Black Park, Exeter, Manchester, Isle of Man, Bowdon, Cambridge, Sherwood Forest, Chester, New Brighton.

The situations where it is found are woods, hedge-sides, gardens, etc.

The perfect insect appears in August.

The caterpillar is dull greyish yellow, with a paler line along the back; the side line whitish, with a row of black spots over it.

The date of the appearance of the caterpillar is in November, and through the winter to April.

It feeds on grass.

The chrysalis is subterranean.

ORTHOSIDÆ.

TRACHEA PINIPERDA.

PINE BEAUTY.

PINE-DEVOURING NOCTUA.

Plate LVI. *Figure* 7.

Localities for this species are York, Langwith, Stockton Forest, Scarborough, Huddersfield, Brighton, Newark? Bowdon, Dunham Park, Wavendon, Manchester, Carlisle, Torwood, and Dunmoor near Stirling, Lyndhurst, Edinburgh, and Birkenhead.

The situations where it is found are pine woods.

The perfect insect appears in March, April, and May; March 29th., April 23rd. and 26th., and May 2nd.

The caterpillar is green, with a white line along the back, another on each side below it of the same colour, as is likewise the side line, below which last-named is an orange one.

The date of the appearance of the caterpillar is in July and August.

It feeds on the fir.

The chrysalis is found below the ground, enclosed in a cocoon of earth and silk.

PACHNOBIA ALPINA.

Plate LVI. *Figure* 8.

Localities for this species are at Cairn Gower, in Perthshire.

The situations where it is found are on high mountains. The perfect insect appears in August.

TÆNIOCAMPA GOTHICA.

HEBREW CHARACTER.

Plate LVI. *Figure* 9.

Localities for this plentiful species are York, Doncaster, Sutton-on-Derwent, Brighton, Bowdon, Manchester, Bere Wood, Epping, Bromsgrove, Rannock, Nunburnholme, etc., etc.

The situations where it is found are hedge-sides, etc.

The perfect insect appears in February, March, and April; February 25th. It has been known in fine condition, even so late as the end of May, and to the 19th. of June.

The caterpillar is green, irrorated with yellowish dots; the line along the back yellowish, the one below it on each side the same; the side line white.

The date of the appearance of the caterpillar is in June.

It feeds on the broom, lilac, clover, etc.

The chrysalis is found enclsoed in a loose cocoon of earth.

This moth often flies in the day-time.

TÆNIOCAMPA LEUCOGRAPHA.

Plate LVI. *Figure* 10.

Localities for this species are York, Wadsworth, Levitt Hag near Doncaster, Darenth Wood, Prestwich, Carlisle,

Henley, Cockermouth, East Grinstead, Leith Hill near Dorking, Horndean, Great Marlow, Barnstaple, and Preston.

The situations where it is found are woods.

The perfect insect appears in February and March, about the middle of the month, and April; February 25, March 10 and 20.

The caterpillar is green, sprinkled with dots of brown and white; the side line rust-coloured, bordered above with black.

The date of the appearance of the caterpillar is in June and July.

It feeds on the plantain (*Plantago lanceolata*).

The chrysalis is found enclosed in a loose earthen cocoon.

TÆNIOCAMPA RUBRICOSA.

RED CHESNUT.

Plate LVII. *Figure* 1.

Localities for this species are York, Doncaster, Sutton-on-Derwent, Bowdon, Lewes, Prestwich, Darenth Wood, Worcester, Burton-on-Trent, Preston, Bristol, Epping Forest, Stowmarket, Rannock, Shrewsbury, Birkenhead, East Grinstead, Torwood, Stirling, Horndean, Plymouth, Kingsbury, Newnham, Manchester, Huddersfield, Darlington, Newark, Lyndhurst, Edinburgh, and Cambridge.

The situations where it is found are woods.

The perfect insect appears at the end of March and in April; March 20.

The caterpillar is dull reddish brown, with a whitish

line along the back, and another of the same colour on each side below it, bordered above with a row of white dots; the side line white.

The date of the appearance of the caterpillar is in June and July.

It feeds on the dock (*Rumex pratensis*).

The chrysalis is enclosed in a loose cocoon of earth.

TÆNIOCAMPA INSTABILIS.

CLOUDED DRAB.

Plate LVII. *Figure* 2.

Localities for this common species are York, Sutton-on-Derwent, Bromsgrove, Brighton, Faversham, Rannock, Bowdon, and Cambridge.

The situations where it is found are fields, etc.

The perfect insect appears in March; March 17.

The caterpillar is green, dotted with black; the line along the back, the one below it, and the side line, yellowish green.

The date of the appearance of the caterpillar is in May and June.

It feeds on the oak, the willow, and the sloe.

The chrysalis is found beneath the ground at the root of trees, and is enclosed in a loose cocoon of earth.

TÆNIOCAMPA OPIMA.

Plate LVII. *Figure* 3.

Localities for this species are York, Doncaster, by the banks of the river Don, Stockton Forest, Langwith, Taunton, Birkenhead, Bristol, Manchester.

The situations where it is found are wooded places.

The perfect insect appears in February, March, and April; February 21—23—24, April 1—13.

The caterpillar is brownish on the upper part, the sides yellowish green, the line on the back, the one below it, and the side line, paler. It is marked subsequently with green stripes.

The date of the appearance of the caterpillar is in June.

It feeds on the poplar and the sallow.

The chrysalis is enclosed in a loose cocoon of earth.

TÆNIOCAMPA POPULETI.

Plate LVII. *Figure* 4.

Localities for this species are York, Sandal Beat near Doncaster, Epping, Dunham Park, Shrewsbury, Manchester, Wimbledon, Preston, Bowdon, Congleton, Black Park, Halton, Kingsbury, Bristol, Birkenhead, Lower Guiting, Cambridge, Bromsgrove, Crewe, Burton-on-Trent, and near London.

The situations where it is found are woods.

The perfect insect appears in March and April; March 17, April 1—13.

The caterpillar is pale whitish green, with a white line along the back, and two narrower ones on the sides.

TÆNIOCAMPA STABILIS.

COMMON QUAKER.

Plate LVII. *Figure* 5.

Localities for this very common species are York, Sutton-on-Derwent, Bromsgrove, Brighton, Faversham, Warter, Bowdon, Cambridge, Rannock, etc., etc.

The situations where it is found are hedge-sides, woods, etc.

The perfect insect appears in March and April.

The caterpillar is pale green, dotted with yellowish, the line along the back yellowish, the one below it hardly traceable, the side line yellowish; there is a yellowish mark on the front of the second segment, and a line of the same colour across the twelfth.

The date of the appearance of the caterpillar is in June and July.

It feeds on the elm and the oak.

The chrysalis is found in a loose earthen cocoon.

TÆNIOCAMPA GRACILIS.

LEAD-COLOURED DRAB.

Plate LVII. *Figure* 6.

Localities for this species are York, Potteric Carr near Doncaster; Sutton-on-Derwent, Birkenhead, Horndean, Warrington, Exeter, Darenth Wood, Bowdon, East Grinstead, Bristol, Darlington, Huddersfield, Cambridge, Epping, Kingsbury, Worcester, Newark, Preston, and near London.

The situations where it is found are woods.

The perfect insect appears in February and March; February 25, March 20.

The caterpillar is rather dark green, with a paler line along the back, and one below it on each side, and a row of pale green dots between them; the side line whitish, edged on its upper part with green.

The date of the appearance of the caterpillar is in May and June.

It feeds on the willow.

The chrysalis is enclosed in a loose cocoon of earth.

This species often flies in the day-time.

TÆNIOCAMPA MINIOSA.

BLOSSOM UNDERWING.

Plate LVII. *Figure 7.*

Localities for this species are Sandal Beat near Doncaster, York, Lewes, Dulwich, Lyndhurst, Horndean, Newnham, Cockermouth.

The situations where it is found are woods.

The perfect insect appears in March and April. April 1—13.

The caterpillar is blue, with a yellow line along the back, and another of the same colour below it on the sides; the side line yellowish, with a row of whitish dots above it.

The date of the appearance of the caterpillar is in May.

It feeds on the oak.

The chrysalis is found enclosed in a loose cocoon of earth.

TÆNIOCAMPA MUNDA.

TWIN-SPOT QUAKER.

Plate LVII. *Figure* 8.

Localities for this species are York, Doncaster, Sutton-on-Derwent, Langwith, Lewes, Epping, Taunton, Newark, Dulwich, Dunham Park, Preston, Worcester, Wimbledon, Bowdon, Stowmarket, Kingsbury, Black Park, Plymouth, Huddersfield, Halton, Marlow, Manchester, Birkenhead, Exeter, Bristol, Arundel, Lyndhurst, Cambridge, East Grinstead, and Burton-on-Trent.

The situations where it is found are woods.

The perfect insect appears in February and March. February 21—23, March 10.

The caterpillar is brown, with a yellowish grey line along the back, with a row of yellowish dots on each side of it; the side line yellowish grey, edged above with dark grey, and beneath it is a white spot on the fifth and sixth segments.

The date of the appearance of the caterpillar is in May and June.

It feeds on the elm and the aspen.

The chrysalis is enclosed in a loose earthen cocoon.

TÆNIOCAMPA CRUDA.

SMALL QUAKER.

Plate LVII. *Figure* 9.

Localities for this common species are York, Warter, Sutton on-Derwent, Langwith, Brighton, Dunham Park, Faversham, Bowdon, Doncaster, Hainault Forest, and Bromsgrove.

The situations where it is found are hedge-sides and woods.

The perfect insect appears in February and March; February 23, March 10.

The caterpillar is pale green, (sometimes greyish or brownish) with a line along the back of whitish green, and another of the same colour on each side below it, a row of dark dots below the two; the side line yellowish, as are the interstices of the segments.

The date of the appearance of the caterpillar is in May.

It feeds on the oak.

The chrysalis is found enclosed in a loose earthen cocoon.

ORTHOSIA SUSPECTA.

Plate LVII. *Figure* 10.

Localities for this species are Askham near York, Doncaster, Stockton Common, Buttercrambe Moor, Storeton and Hargrave Hall near Birkenhead, Worksop, Carlisle, Edinburgh, Huddersfield, Sherwood Forest, Manchester.

The situations where it is found are woods.

The perfect insect appears in July and August.

The chrysalis is found enclosed in a cocoon below the earth.

ORTHOSIA YPSILON.

DINGY SHEARS.

Plate LVII. *Figure* 11.

Localities for this neat species are York, Askham Bog, Nunburnholme, Lewes, Stowmarket, Kingsbury, Exeter, Cambridge, Burton-on-Trent, Bristol, and Birkenhead.

The situations where it is found are marshy places.

The perfect insect appears in July.

The caterpillar is blackish brown, with a broad paler stripe along the back, between which is a chain-shaped streak; the side line dull yellowish grey.

The date of the appearance of the caterpillar is in June.

It feeds on the willow and the poplar.

The chrysalis is found below the surface, enclosed in a cocoon.

ORTHOSIA LOTA.

RED-LINED QUAKER.

Plate LVIII. *Figure* 1.

Localities for this rather common and widely-distributed species are York, Nunburnholme, Sutton-on-Derwent, Bromsgrove, Warter, Brighton, Bowdon, Faversham,

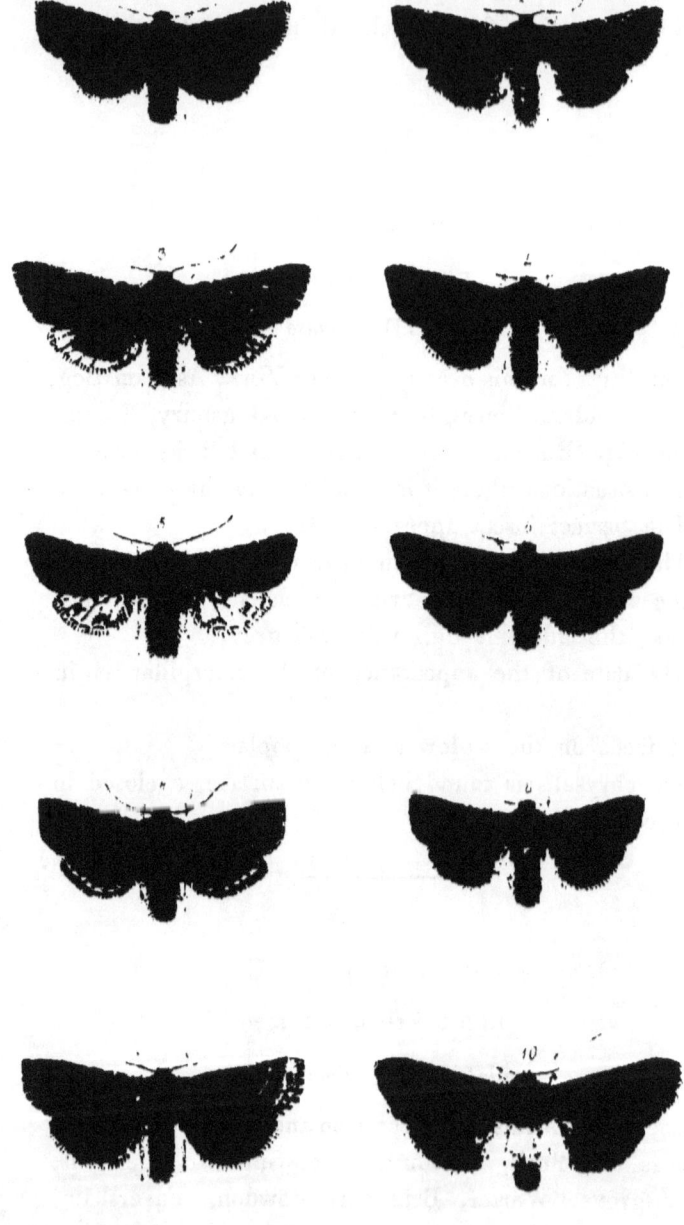

Lewisham, Marlow, the New Forest, Barnstaple, Ely, Peterborough, etc., etc.

The situations where it is found are gardens, hedgesides, etc.

The perfect insect appears in September and October.

The caterpillar is greyish brown, with a tinge of violet, and dotted with white; the line along the back white, but interrupted.

The date of the appearance of the caterpillar is in May and June.

It feeds on the willow.

The chrysalis is found below the surface, in a cocoon.

ORTHOSIA MACILENTA.

YELLOW-LINED QUAKER.

Plate LVIII. *Figure 2.*

Localities for this species are York, Nunburnholme, Sutton-on-Derwent, Brighton, Birkenhead, Stowmarket, Lewes, Marlow, Bowdon, Bristol, Arundel, Cambridge, Kingsbury, Wavendon, Barnstaple, Lower Guiting, Ely, Lyndhurst, Darlington, Huddersfield, Shrewsbury, Manchester.

The situations where it is found are gardens, woods, etc.

The perfect insect appears in September and October.

The caterpillar is reddish brown, with numerous white dots; the line along the back white; the one below it on each side also white, and the side line white; the head very small.

The date of the appearance of the caterpillar is in September and October.

It feeds on the beech.

The chrysalis is found under the ground, enclosed in a cocoon.

ANCHOCELIS RUFINA.

Plate LVIII. *Figure* 3.

Localities for this handsome species are Sutton-on-Derwent, and Crompton's Coppice near York, Hertford, Lyndhurst, Burton-on-Trent, Brighton, Manchester, Cambridge, Birkenhead, Edinburgh, Lewes, Plymouth, Lower Guiting, Bristol, Scarborough, Exeter, Wavendon, Barnstaple, Ipswich, Torwood, Stirling, Halton, Tenterden, Huddersfield, Worcester, and near London.

The situations where it is found are woods.

The perfect insect appears in September.

The caterpillar is orange coloured, with a paler line along the back, and a row of whitish dots on each side; the side line white.

The date of the appearance of the caterpillar is in May.

It feeds on the oak.

The chrysalis is subterranean.

ANCHOCELIS PISTACINA.

PALE-HEADED CHESNUT.

Plate LVIII. *Figure* 4.

Localities for this species are York, Hanging-Heaton near Dewsbury, Faversham, Brighton, Bromsgrove, Birkenhead, Exeter, Nunburnholme, Bristol, Halton, Lewes, Lewisham, Cambridge, Kingsbury, Chichester, Lower Guiting, Lyndhurst, Sidmouth, Scarborough, Wavendon, Worcester, Barnstaple, Weston-Super-Mare, Shrewsbury, Tenterden.

The situations where it is found are hedge-sides, gardens, etc.

The perfect insect appears in September and October.

The caterpillar is green, or dull yellowish, with spots of white; the line along the back darker, as is the one above it on the sides; the side line whitish.

The date of the appearance of the caterpillar is in May and June.

It feeds on the dock (*Rumex pratensis*).

The chrysalis is found below the earth.

ANCHOCELIS LUNOSA.

LUNAR UNDERWING.

Plate LVIII. *Figure* 5.

Localities for this species are Brighton, Faversham, Bowdon, Birkenhead, Darenth Wood, Worcester, Scarborough, Lewisham, Wavendon, Plymouth, Chichester, Lewes, Tenterden, Lyndhurst, Kingsbury, Barnstaple, Stowmarket, Exeter, Darlington, Shrewsbury, Bristol, Edinburgh, Cambridge.

The situations where it is found are woods.

The perfect insect appears in September.

The caterpillar is green, or dull greyish green, with a

line along the back, and one below it on the sides of whitish; the side line, which is narrower, is white and bordered on its upper edge with black; the spots black.

The date of the appearance of the caterpillar is in April.

It feeds on grass.

The chrysalis is subterranean.

ANCHOCELIS LITURA.

BROWN-SPOT PINION.

Plate LVIII. *Figure* 6.

Localities for this common species are York, Huddersfield, Nunburnholme, Bromsgrove, Hanging Heaton near Dewsbury, Brighton, Carron, Stirling, Bowdon, Birch Wood, Darenth Wood, Faversham, New Forest, Arundel, Peterborough.

The situations where it is found are woods, hedgesides, gardens, etc.

The perfect insect appears in September and October.

The caterpillar is green or pale brown, with a paler line along the back and another below it on either side; the side line whitish; the spots yellowish.

The date of the appearance of the caterpillar is in June and July.

It feeds on various low plants.

The chrysalis is subterranean.

GLÆA VACCINII.

CHESNUT.

Plate LVIII. *Figure* 7.

Localities for this also common species are Doncaster, Crompton's Coppice near York, Brighton, Bromsgrove, Bowdon, Lewisham, Epping, Plymouth, Weston-super-Mare, Barnstaple, Torwood, Stirling.

The situations where it is found are woods, hedge-sides, gardens, etc.

The perfect insect appears in October and November, and continues till February and March; February 23.

The caterpillar is dark brown, with a paler line along the back and another on each side of it; the spots pale grey; the side line dull yellowish grey.

The date of the appearance of the caterpillar is in June and July.

It feeds on the oak; also on low plants.

The chrysalis is found below the earth.

GLÆA SPADICEA.

BLACK CHESNUT.

Plate LVIII. *Figure* 8.

Localities for this species are Sutton-on-Derwent and Crompton's Coppice near York, Manchester, Bromsgrove, Nunburnholme, Brighton, Shrewsbury, Plymouth, Stowmarket, Bowdon, Birkenhead, Bristol, Barnstaple, Burton-on-Trent, Cambridge, Halton, Tenterden, Lower Guiting,

Darlington, Huddersfield, Wavendon, Lewisham, Kingsbury, Lewes, Lyndhurst, Worthing, Worcester.

The situations where it is found are woods, hedgesides, gardens, etc.

The perfect insect appears in October and November, and so through the winter to February and March.

The caterpillar is dark brown, with a paler line along the back and another on each side below it; the side line also paler brown, the intermediate space blackish brown.

The date of the appearance of the caterpillar is in May and June.

It feeds on the sloe, the hawthorn, and low plants.

The chrysalis is found below the surface.

GLÆA ERYTHROCEPHALA.

Plate LVIII. *Figure* 9.

Localities for this species are Hurst near Brighton, Lewes, Marlow, Plymouth, Weston-super-Mare.

The perfect insect appears in October; October 20, and November.

The caterpillar is brownish grey dotted with white, the second segment with a black patch in which are two white lines.

The date of the appearance of the caterpillar is in May.

It feeds on various low plants.

The chrysalis is subterranean.

SCOPELOSOMA SATELLITIA.

SATELLITE.

Plate LVIII. *Figure* 10.

Localities for this common species are York, Doncaster, Sutton-on-Derwent, Nunburnholme, Hanging Heaton, Bromsgrove, Bowdon, Weston-Super-Mare, Faversham, Epping, The New Forest, Exeter, Plymouth, Durham, Torwood, Stirling, Brighton, Barnstaple, Lewisham.

The situations where it is found are woods, hedge-sides, gardens, etc.

The perfect insect appears in October and November, and continues till February, March, and even April; February 23.

The caterpillar is dark blackish brown, with three white lines on the second segment, and a white spot on the second, third, fourth, fifth, and twelfth segments underneath the side line.

The date of the appearance of the caterpillar is in May and June.

It feeds on the oak, the beech, the elm, etc.

DASYCAMPA RUBIGINEA.

DOTTED CHESNUT.

Plate LIX. *Figure* 1.

Localities for this rare species are Bromsgrove, where I took one myself, when at the Grammar School of King Edward the Sixth there, Norbury Park, Brighton, Worcester, Arundel, Newnham, Lyndhurst, Bristol, Freshwater, Gloucester, Mickleham, Barnstaple, Dublin, Weston-Super-Mare, Plymouth, Marlow, Exeter.

The situations where it is found are hedge-sides.

The perfect insect appears in October and November, and lives through the winter to March and April; October 24.

The caterpillar is yellowish brown, with a blackish spot on the back of each segment.

The date of the appearance of the caterpillar is in June and July.

It feeds on the oak and several plants.

The chrysalis is found in a loose cocoon mixed with earth.

I have to thank Mr. Allis, of York, for his obliging loan of the specimen from which the plate is taken.

HOPORINA CROCEAGO.

ORANGE UPPER WING.

Plate LIX. *Figure* 2.

Localities for this species are Selling Wood near Faversham, Dartmoor, Brighton, Lewes, Darenth Wood, Lyndhurst, Black Park, Tenterden, East Grinstead, Worcester, and Barnstaple.

The situations where it is found are oak woods.

The perfect insect appears in October and November, and continues through the winter till March and April.

The caterpillar is reddish yellow, with a row of slanting blackish lines on each side of the back on the fifth, sixth, seventh, eighth, ninth, tenth, eleventh, and twelfth segments, meeting on the middle of the back.

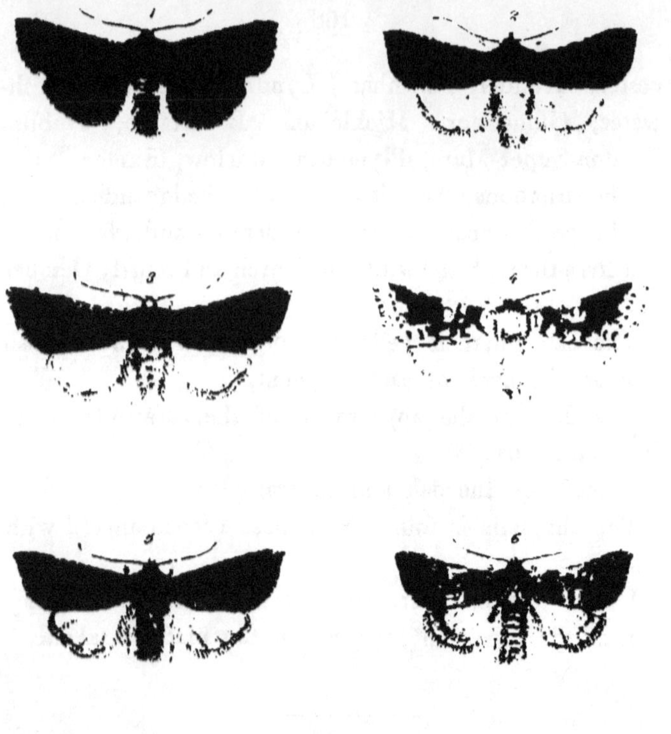

The date of the appearance of the caterpillar is in May and June.

It feeds on the oak.

The chrysalis is found below the surface.

XANTHIA CITRAGO.

ORANGE.

Plate LIX. *Figure* 3.

Localities for this species are York, Walthamstow, Epping Forest, Dartmoor, Lewes, Worcester, Manchester, Huddersfield, Stowmarket, Halton, Shrewsbury, Exeter, Bristol, Kingsbury, and Birkenhead.

The perfect insect appears in September; September 1—11.

The caterpillar is grey, with a whitish line along the back, and another of the same colour on each side of it, the latter edged on each segment with a black spot and three or four white dots; the side line whitish edged above with black.

The date of the appearance of the caterpillar is in May and June.

It feeds on the lime.

The chsysalis is found below the earth.

XANTHIA CERAGO.

Plate LIX. *Figure* 4.

Localities for this species are York, Sutton-on-Derwent,

Langwith, Askham Bog, Faversham, Selling, Edinburgh, Lewes, Darlington, Plymouth, Bowdon, Lower Guiting, Shrewsbury, Ipswich, Cambridge, Bristol, Scarborough, Burton-on-Trent, Southport, Worthing, Birkenhead, Tenterden, Manchester, Lyndhurst, Stowmarket, Halton, Huddersfield.

The situations where it is found are woods.

The perfect insect appears in September.

The caterpillar is of a violet brown colour, with a line along the back bordered by two pale ones; the side line greyish.

The date of the appearance of the caterpillar is in April and May.

It feeds on the catkins of the sallow, and afterwards on low plants.

The chrysalis is found below the earth.

XANTHIA SILAGO.

PINK-BARRED SALLOW.

Plate LIX. *Figure* 5.

Localities for this species are York, Sutton-on-Derwent, Askham Bog, Langwith, Faversham, Selling, Cambridge, Lewes, Exeter, Chichester, Bowdon, Worthing, Burton-on-Trent, Edinburgh, Ipswich, Tenterden, Stowmarket, Bristol, Darlington, Shrewsbury, Manchester, Birkenhead, Lower Guiting, Scarborough, Huddersfield, and Halton.

The situations where it is found are woods and moist places.

The perfect insect appears in September.

The caterpillar is reddish brown, with numerous dots of

brown, red, yellow, and white, of which latter the side line is formed.

The date of the appearance of the caterpillar is in June.

It feeds on several low plants.

The chrysalis is subterranean.

XANTHIA AURAGO.

BARRED SALLOW.

Plate LIX. *Figure 6.*

Localities for this species are York, Lydiate near Liverpool, Brighton, Freshwater, Birch Wood, Ipswich, Darenth Wood, Bristol, Stowmarket, Black Park, Halton, Worcester, Henley-on-Thames, Marlow, and Arundel.

The situations where it is found are woods.

The perfect insect appears in September.

The caterpillar is grey with darker slanting streaks.

The date of the appearance of the caterpillar is in May.

It feeds on the beech.

The chrysalis is found below the surface.

XANTHIA GILVAGO.

Plate LIX. *Figure 7.*

Localities for this species are York, Rotherham, Doncaster, Burton-on-Trent, Freshwater in the Isle of Wight, Ipswich, Cambridge, Derby, Deal, Shrewsbury, Brighton, and Bourne in Lincolnshire.

The perfect insect appears in September.

It feeds on the seeds of the elm.
The chrysalis is found below the surface.

XANTHIA FERRUGINEA.

BRICK-COLOURED MOTH.

Plate LIX. *Figure* 8.

Localities for this common species are York, Nunburnholme, Sutton-on-Derwent, Coombe Wood, Faversham, The New Forest, Brighton, Bowdon, and Plymouth.

The situations where it is found are gardens, woods, etc.

The perfect insect appears in September and October.

The caterpillar is pale reddish brown, with numerous small darker spots; the line along the back paler, most distinct on the hinder segments; the side line pale.

The date of the appearance of the caterpillar is in April and May.

It feeds on the young buds of the poplar.

The chrysalis is found underneath the ground.

CIRRÆDIA XERAMBELINA.

CENTRE-BARRED SALLOW.

Plate LIX. *Figure* 9.

Localities for this rare species are York, and near Howsham, where my brother Beverley R. Morris, Esq. took one, Scarborough, Worcester, Bromsgrove, where I, or one of my schoolfellows, once took one, Halton, Lewisham, Stowmarket, Darlington, Bristol, Cambridge, Ipswich, and Burton-on-Trent.

The situations where it is found are hedge-sides.

The perfect insect appears in September.

The caterpillar is marbled greyish and brown, with a pale line on each side of the back edged above with darker; the side line paler, interrupted, and edged with black.

The date of the appearance of the caterpillar is in June and July.

It feeds on the ash.

The chrysalis is found at the roots of trees below the earth.

COSMIDÆ.

TETHEA SUBTUSA.

OLIVE MOTH.

Plate LX. Figure 1.

Localities for this species are York, Sutton-on-Derwent, Kingsbury, Bristol, Lewes, Carlisle, Bowdon, Huddersfield, Exeter, Worcester, Stowmarket, Burton-on-Trent, Manchester, Cambridge, Halton, Birkenhead, Clapham, and Hammersmith near London.

The situations where it is found are woods.

The perfect insect appears in July and August.

The caterpillar is yellowish green, with a pale yellow line along the back, and another of the same colour on either side of it; the side line pale yellowish; the head whitish green, the mouth black.

The date of the appearance of the caterpillar is in April and May.

It feeds on the aspen and the poplar.

TETHEA RETUSA.

DOUBLE KIDNEY.

Plate LX. *Figure* 2.

Localities for this species are York, Brighton, Bowdon, Cambridge, Black Park, Tenterden, Arundel, Wavendon, and Worcester.

The perfect insect appears in August and September.

The caterpillar is green, with a whitish line along the back, and another of the same colour on either side of it; the side line also whitish.

The date of the appearance of the caterpillar is in May.

It feeds on the sallow and the poplar.

EUPERIA FULVAGO.

ANGLE STRIPED SALLOW.

Plate LX. *Figure* 3.

Localities for this species are Carlisle, Sherwood Forest, and Lewisham.

The situations where it is found are birch woods.

The perfect insect appears in August.

The caterpillar is pale green or greyish, with a white line along the back, and another of the same colour on either side of it; the side line whitish, edged above with black; the spots whitish.

The date of the appearance of the caterpillar is in June.

It feeds on the oak and the birch.

The chrysalis is found on the surface of the earth

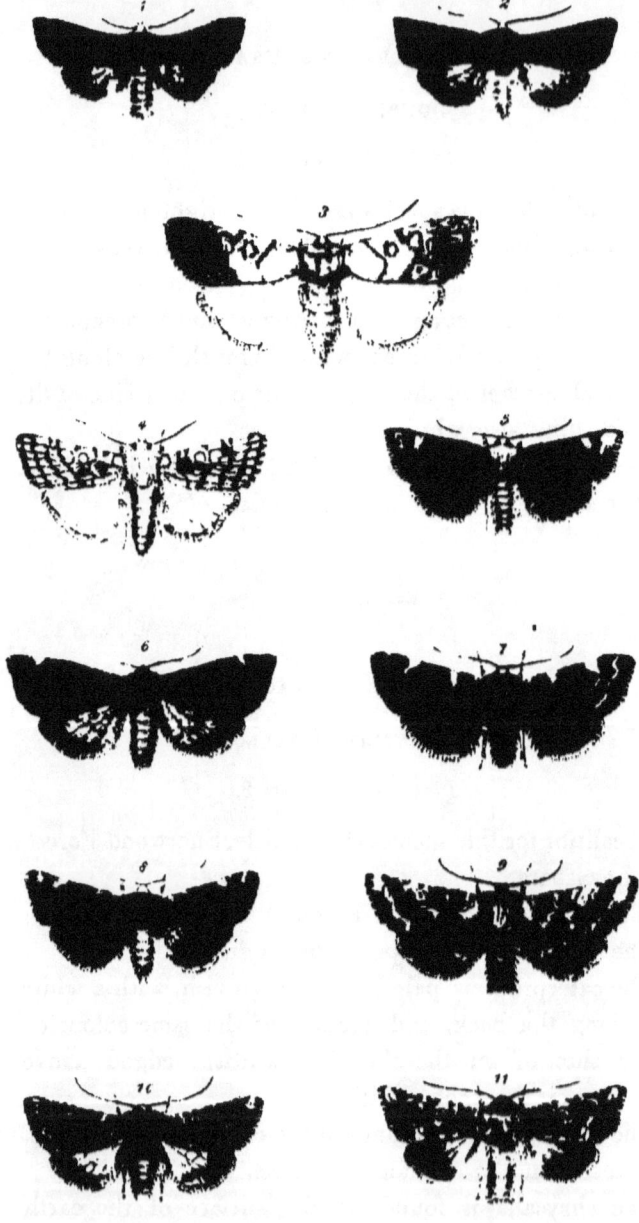

Plate VI.

enclosed in an earthen cocoon, and covered with a violet-coloured dust.

DICYCLA Oo.

SCALLOP-WINGED OAK MOTH.

Plate LX. *Figure* 4.

Localities for this species are Black Park, Lyndhurst, Worcester, Tenterden, and Stowmarket.

The perfect insect appears in June.

The caterpillar is brownish black, with a very distinct white line along the back, another of the same colour on each side of it, and the side line also smaller.

The date of the appearance of the caterpillar is in May and June.

It feeds on the oak.

The chrysalis is found on the surface of the ground, enclosed in a cocoon.

COSMIA TRAPETZINA.

DUN BAR.

Plate LX. *Figure* 5.

Localities for this species are Langwith near York, and Sutton-on-Derwent, Brighton, Charmouth, etc., etc.

The situations where it is found are woods.

The perfect insect appears in July and August.

The caterpillar is greenish, with a white line along the

back, another on either side of it, and the side line, of the same colour; the spots black or dark green.

The date of the appearance of the caterpillar is in May and June.

It feeds on the oak, the birch, etc.

The chrysalis is found between leaves, or in a cocoon on the surface of the earth.

COSMIA PYRALINA.

LUNAR-SPOTTED PINION.

Plate LX. Figure 6.

Localities for this species are Birch Wood, Epping, Swinhope, Barham, Brighton, Shrewsbury, Black Park, Stowmarket, Monks' Wood, Tenterden, Worcester.

The situations where it is found are woods.

The perfect insect appears in August.

The caterpillar is pale green, with a paler line along the back, and another of the like colour on each side of it; the side line yellowish edged above with black; the spots yellowish green; the head dark green.

The date of the appearance of the caterpillar is in April and May.

It feeds on the plum and the pear.

The chrysalis is found between leaves, or in a cocoon on the surface of the ground.

To Mr. Allis, of York, I am also here much obliged for having lent the specimen from which the engraving is taken

COSMIA DIFFINIS.

WHITE PINION-SHOT.

Plate LX. *Figure* 7.

Localities for this species are Doncaster, Bromsgrove, Cambridge, Rainhill near Liverpool, Brighton, Worcester, Black Park, Reigate, Lewes, Lewisham, Worthing, Plymouth, Kingsbury, Arundel, Wavendon, Halton, Exeter, Cambridge, Dorking, Stowmarket, Burton-on-Trent, and Bristol.

The situations where it is found are woods, hedge-sides, gardens, etc.

The perfect insect appears in July and August.

The caterpillar is pale green, with the line along the back, and the one on each side of it whitish; the side line pale yellow; the spots black in a rim of white; the head black.

The date of the appearance of the caterpillar is in May.

It feeds on the elm.

The chrysalis is found between leaves, or in a cocoon on the ground.

COSMIA AFFINIS.

LESSER PINION-SPOT.   BROWN PINION-SPOT.

Plate LX. *Figure* 8.

Localities for this species are Doncaster, Bromsgrove, Lewes, Cambridge, Worthing, Hightown, and Sefton near Liverpool, Black Park, Reigate, Worcester, Lewisham,

Wavendon, Plymouth, Arundel, Stowmarket, Lyndhurst, Dorking, Shrewsbury, Kingsbury, Halton, Exeter, Bristol.

The situations where it is found are hedge-sides, woods, etc.

The perfect insect appears in July and August.

The caterpillar is blackish green, with the line along the back white and broad; the line below it on each side white and narrow; the side line white and narrow, the head pale green.

The date of the appearance of the caterpillar is in May.

It feeds on the elm.

The chrysalis is found between leaves, or in a cocoon on the earth.

HADENIDÆ.

EREMOBIA OCHROLEUCA.

DUSKY SALLOW.

Plate LX. *Figure 9.*

Localities for this species are York, Doncaster, Deal, Sheldwick, Brighton, Faversham, Bristol, Stowmarket, Ventnor, Lewes, and Peterborough.

The situations where it is found are heathy downs and waste places.

The perfect insect appears in July and August.

The caterpillar is yellowish green; the side line pale yellow; the spots black.

The date of the appearance of the caterpillar is in May and June.

It feeds on grasses of different sorts.

The chrysalis is subterranean.

DIANTHÆCIA CARPOPHAGA.

TAWNY SHEARS.

Plate LX. *Figure* 10.

Localities for this species are York, Sutton-on-Derwent, Stockton Forest, Brighton, Faversham, Sudbury, Ventnor, Bristol, Halton, Lewes, Edinburgh, Cambridge, Mickleham, and near London.

The perfect insect appears in June and July.

The caterpillar is dark grey, with the line along the back whitish and broad, the line below it on each side pale grey; the side line pale grey; the head reddish, with two dark brown lines.

The date of the appearance of the caterpillar is in August.

It feeds on the bladder campion (*Silene inflata*).

DIANTHÆCIA CAPSOPHILA.

Plate LX. *Figure* 11.

Localities for this species are near Dublin.

The situations where it is found are rocks on the sea coast.

The perfect insect appears in June and July.

The date of the appearance of the caterpillar is in July and August.

It feeds on the seed vessels of the sea campion, (*Silene maritima.*)

The chrysalis is found at the root of the plant, enclosed in a cocoon of earth.

This moth is attracted by light.

I have to thank Edwin Birchall, Esq., of Oakfield Villa, near Birkenhead, for the above information about this species, and T. H. Allis, Esq., of York, for the loan of the specimen from which the figure on the plate is engraved.

END OF VOLUME II.

www.ingramcontent.com/pod-product-compliance
Lightning Source LLC
Chambersburg PA
CBHW020809230426
43666CB00007B/929